The Alien Colonization of Earth's Waterways

A Reference Guide to UFO/USO Water-Related Activity

By Debbie Ziegelmeyer

The Alien Colonization of Earth's Waterways
A Reference Guide to UFO/USO Water-Related Activity

By Debbie Ziegelmeyer

©Copyright 2021 by Debbie Ziegelmeyer
Photos copyrighted by Debbie Ziegelmeyer unless otherwise noted
Editor: Margie Kay

ISBN 978-1-7378996-2-4

Black and white version

Published in the United States of America

Publisher:

UnX Media
PO Box 1166
Independence, Missouri 64051
www.unxmedia.com

Contact the author through UnX Media at
unxnewseditor@gmail.com

Dedication

I wish to thank my husband Wayne who has supported me not only in the writing of this book, but behind the scenes for over twenty-five years while I traveled across the country investigating the UFO phenomenon. Christmas, birthdays, Mother's Day, our wedding anniversary, and sometimes just because, gifts from Wayne have always had an investigating theme. Over the years he has given me everything from camera equipment with the latest night vision available to the public, night vision binoculars, EMF meters, a range finder, a surface and an underwater metal detector, to the latest scuba diving gear. Wayne has helped me dress up three different Jeep Wranglers over the years with LED lights, three-inch lifts, skid plates, large wheels, and tires, and finally my latest gifts this past Christmas and birthday which were a new tube front bumper and a wench for my current Jeep. Wayne has dropped me off at the airport in the early hours of the morning and has picked me up sometimes near or after midnight. It is not an easy task traveling to and from St. Louis Lambert Field which is a 40-minute drive from our home, and then go on to work or wake up the next morning with only a few hours of sleep. Wayne always sits up waiting for me to get safely home from investigations, speaking engagements, and conferences and has never once complained. Without my husband I am sure I never could have accomplished as much as I have and kept this up for so many years. When we were married right out of high-school I can be quite sure he never imagined I would someday lead him down this long unbelievable road, but he has hung in there all the way.

Thank you, Wayne, for your love and support especially over the past 25 years. I have never taken anything you have done in support of me and my interests for granted, and I will love you for this and more 'til the end of time and then some.

Acknowledgements

Thank you to Margie Kay who encouraged me to write this book and patiently guided me through the process.

In remembrance of Carl Feint who I had the pleasure of speaking with and exchanging theories on water related UFO sightings. Cheers to you Carl, you are missed.

My appreciation to John Schuessler who, for years, has sent me every article he finds which is related to water UFO/USO sightings. John sent me large envelopes with newspaper articles, magazine articles, and more recently, forwarded website links. Each envelope I opened contained a new piece of uncovered secrets which to me were a form of underwater treasure. Thank you, John.

Thank you to John Greenewald, Jr. with The Black Vault, for allowing me to use some of his Freedom of Information Act (FOIA) findings.

Contents

Introduction

What deep water secrets are the oceans hiding? How much do we actually know about our planet? Two very interesting questions which I will attempt to answer in the pages to follow. Seventy percent of the Earth is covered by water and according to statistics, one in five UFO sightings are water related. The National Oceanic and Atmospheric Association (NOAA) states that "using increasingly sophisticated tools, technologies, and sensors, we learn more about our oceans every day; however, we have still explored less than five percent of this vast underwater frontier."

Before we speculate the existence of other worldly beings living beneath the surface of our oceans, seas, rivers, and lakes, we need to understand when humans first began exploring the mysteries underwater and what technologies humans were capable of in the past and in the present. Knowing where the deepest parts of the oceans and major trenches are located is also vital in determining possibilities for the existence of deep underwater bases and their possible or probable locations and occupants.

The investigation has to begin with the history of early submarines and follow through to today's U.S. Navy with its four classes of navy submarines. We then move on to tunnels such as those which go under the Hudson River in New York and the English Channel in Europe. There are also underwater research bases to consider which have been in existence since the 1960's. Research into and discussions of under-ocean and in-bottom military bases began decades ago. In 1968 the Stanford Research Institute discussed the construction of dozens of undersea bases. The study was titled "Feasibility of Manned In-Bottom Bases."

We know humans can only survive underwater at depth for minutes without a pressurized habitat and can only survive in the vacuum of space in a pressurized craft, suit, or space station. Why, in this unfriendly environment are alien craft using our oceans, seas, and other waterways? Are alien craft using the hydrogen in the water to replenish their electromagnetic field? Are they also using our waterways to hide or establish bases on Earth? How long have other worldly beings been visiting the Earth and are they just visiting or are they colonizing our planet and have been for centuries.

Research has told us through witness reports, pictures, and more recently video including Navy pilot video, that unknown unidentified craft have been sighted over and coming in and out of our water ways for centuries. Several researchers-investigators such as Vladimir Azhazha, Dr. Virgilio Sanchez-Ocejo, Lt. Col. Wendelle C. Stevens, Preston Dennett, Carl W. Feindt, Maximillien de Lafayette, Jorge Martin, Ivan Sanderson, Richard Sauder, Paul Stonehill, Phillip Mantle, and many others including myself, have accumulated thousands of hours of research, documents, witness interviews, police and military reports, photos, and videos documenting the existence of not only underwater military bases but the possibility of "alien being bases.". Organizations such as MUFON (Mutual UFO Network), NUFORC (National UFO Recording Center), CUFOS, and many others have been collecting and some such as MUFON, have been investigating witness sighting reports for over fifty years. Witness reports have been obtained of abductees being taken underwater to dry base-like facilities which I believe is enough evidence to make a legitimate case. Many of the first UFO (unidentified flying object) and

USO (unidentified submerged object) reports were documented in ship logbooks long before humans were capable of flight or underwater exploration. There are multiple documents available discussing underwater bases and the planning of underwater facilities dating back to the 1960's. Some of these are known research centers and some are thought to be underwater military bases. But, in some areas, massive amounts of UFO sightings and activity are being reported near, over and coming in and out of bodies of water near these facilities.

Any subject matter involving the field of UFO investigation is based on evidence available and speculation to fill the gaps. This research covers a brief history of human underwater research, adventures, and military involvement past and present. It also includes water related UFO sighting reports, some dating back hundreds of years which was obviously long before humans were safely capable of navigating beneath our oceans, seas, and waterways. The possibility of underwater "alien life" bases or habitats is a very strong possibility based on this research and that of many others in the field of ufology. Many governments have begun releasing documents to the public of encounters with unknown, "non-Earth" craft. Most recently, documents and video have been released from the secret archives of both Russian and the U.S. some under the F.O.I.A. (Freedom of Information Act) in the U.S. and others by those within the governments who believe the public is finally entitled to the truth. Or is it that the existence of "alien life" on Earth is becoming harder to conceal, and what have been referred to as "visitors" and "explorers" in the past, are in reality RESIDENTS! Christopher Columbus was an explorer whose mission was colonization of the "new world". The definition of resident, Webster's Dictionary: ("living in a place for some length of time: residing.") These residents who have been secretly inhabiting our planet for centuries need to be exposed.

Chapter 1

Ancient UFO Water Sightings

There are numerous reports which have been documented of unidentified or unexplained phenomenon going back centuries across the world. These have been written not only in several languages but depicted in pictographs on ancient monuments, native American and Mayan mountain faces and on cave walls which have been discovered worldwide. The following are witness accounts of documented ancient sightings of unidentified craft and "beings" seen in the vicinity of bodies of water.

In the year A.D. 1067 from Geoffrey Gaimar's Lestoire des English:

"In this year people saw a fire that flamed and burned fiercely in the sky. It came near the earth and for a little time brilliantly lit it up. Afterwards, it revolved, ascended on high, and descended into the sea."

Jacques Vallee's 1969 book "Passport to Magonia"

This book documents the following sighting: In 1361, a flying object described as being "shaped like a drum, about twenty feet in diameter" emerged from the inland sea off western Japan. The location was listed in the captain's logbook as in the Sea of Japan at latitude of 40-00 N and Longitude of 135-00 E.

Christopher Columbus

While patrolling the deck of the Santa Maria at about 10:00 PM on October 11, 1492, Christopher Columbus thought he saw "a light glimmering at a great distance." He hurriedly summoned Pedro Gutierrez, "a gentleman of the king's bedchamber," who also saw this light. Columbus described the light as "a small wax candle that rose and lifted up, which too few seemed to be an indication of land." His son Ferdinand also described the object as a candle which went up and down. After a short time, the object vanished from sight only to reappear several times throughout the night, each time dancing up and down "in sudden and passing gleams." The mysterious light which was seen four hours before land was sighted, was never explained. Queen Isabella of Spain presented Columbus the "Royal Reward" for his reported sighting. The unexplained light was reported in Columbus' journal, Ferdinand Columbus' Vita del Ammiraglio (The Life of the Admiral), the proceedings of the Pleitos Colombinos (which was the long lawsuit involving the heirs of Columbus) and additional sources.

The Madonna with Saint Giovannino

Figure 1 The Madonna with Saint Giovannino by Domenico Ghirlandaio Image: Wikimedia Commons

This famous painting circuit late 1400's, shows an unidentified craft hovering over the water in the background behind the left shoulder of the Madonna. A man and his barking dog seem to be staring up at this object from the shore. Rays of light appear around the craft in the background as the Madonna in the foreground seems to be praying over her frightened children.

Matha, France

In 1586 just outside the small French village of Matha in the Charente-Maritime, near Angoulème, north of Bordeaux, a "horrid" cloud appeared in the sky that resembled a round, brown hat. The incident was witnessed by the French poet Théodore Agrippa d'Aubigné (1552–1630) and the Marquis de Tors:13. The Marquis, who was lord of the place, took his guest to a garden, shortly before sunset. They saw a round cloud come down over the hamlet of Beauvais-surMatha,

with a color that was "horrid to see", for which one is forced to use a Latin word: subfusca (dark brown). This cloud resembled a hat with an ear in the middle, the color of the throat of an Indian rooster. The hat with its sinister sign came into the steeple and melted there. It appeared to be 9.32 miles (15 km) away, no larger than the moon as seen by the human eye.

Barhofft, Sweden

By Tara MacIsaac , Epoch Times
On April 8, 1665, around 2 p.m., fishermen anchored near Barhöfft (then in Sweden, now in Germany) reported seeing ships in the sky battling each other. After the battle, a dark object hovered in the sky. "After a while out of the sky came a flat round form, like a plate, looking like the big hat of a man... Its color was that of the darkening moon, and it hovered right over the Church of St. Nicolai. There it remained stationary until the evening. The fishermen, worried to death, didn't want to look further at the spectacle and buried their faces in their hands. On the following days, they fell sick with trembling all over and pain in head and limbs. Many scholarly people thought a lot about that," wrote Erasmus Francisci in "Der wunder-reiche Ueberzug unserer Nider-Welt/Oder Erd-umgebende" in 1689. Francisci had gathered news reports from 1665 related to the event. The "scholarly people" who considered the event and the illness could not discern the causes. In the June 2015 edition of EdgeScience magazine , Chris Aubeck and Martin Shough detail their investigation of the event. Aubeck is the founder of the historical research group Magonia Exchange, an international archival project, and a prolific writer on the subject of UFOs as cultural history. Shough is a research associate for the National Aviation Reporting Center on Anomalous Phenomena (NARCAP). The event was first recorded in Leipzig in 1665 and then retold in 1671 by

Johannes Schefferus (1621–1679) in his book Memorabilium Sueticae Gentis Exemplorum. The latter was a source for the German polymath Erasmus Francisci (a.k.a. Erasmus Finx, 1627–1694).

Japan, February 22, 1803, The Utsuro-Bune Legend

As told in the book "Toen shosetsu", the incident occurred on February 22, 1803, when local fishermen were working as they did daily along the shore in Harayadori in the Hitachi province. The fishermen noticed what was described as a "bizarre and extraordinary object" floating on the water, moving in their direction under no apparent control. Fascinated by this, "boat-like vessel" the fishermen met the boat part way to shore towing it in with them.

The fishermen estimated the vessel to be approximately 10 feet high and almost twice as wide. It was split into a top half and bottom half, with the top appearing to be made of a red wood-like material. The bottom of the vessel was much more metallic and solid with the overall shape being almost identical to an incense burner. The windows were covered with strange bars of an unknown material and were covered over with a fluid-like substance. When they looked inside through the windows which appeared to be made of some kind of clear crystal, they saw an interior that appeared similar to a wooden rice bowl. Utsuro translates to hollow, and bune translates to boat or hollow boat. These details described by the fishermen is where the event's name "Utsuro-Bune" originated. This shape description is very similar to what we refer to in modern times as a "flying saucer".

The inside of the vessel appeared to have strange foreign writing on the walls. The fishermen also found a bottle of water, something similar to a cake, a supply of meat, and two large pieces of material which they believed to be bed sheets. Moments later, the fishermen noticed a beautiful young woman who appeared to be around 20 years old. The women's hair, and eyebrows, were red and where her natural hair finished there appeared to be some type of extensions, possibly made of white fur. Her skin had a peculiar "pink" tone to it, much lighter than their own. The clothes she wore were long and smooth and made of a material unknown to the fishermen. The beautiful young mysterious woman was holding a light-colored box that she held on to tightly and would not allow any of the villagers to touch. When she spoke, it was in a language they did not understand or were familiar with.

With the foreign woman safely ashore, the fishermen, nervously pushed the strange boat back out to sea letting the tide take it. In some versions of the tale the fishermen place the woman back in the boat when they push it back out to sea. Other versions tell of how the villagers allowed the woman to remain there with them until she grew old. The strange box that the woman grasped so tightly not allowing them to touch or get near, is similar to stories written about UFO encounters and aliens who carry strange "black boxes" which are sometimes fatal to humans who touch them.

In a legend dating back to the 600s, a tale speaks of another fisherman called Wakegoro who would make a similar discovery of a beautiful young girl in a vessel with the description being almost identical to that described in the Utsuro-Bune Legend. The girl from the 600's could understand the fishermen, and could speak his language, the girl from the Utsuro-Bune Legend could not. This young girl claimed to be the daughter of the emperor of China, but due to the dislike of her stepmother, had been forced to go into hiding for her own safety. The fishermen, according to legend, would take care of the young girl raising her as his own. This legend is part of the origin myths of the Kono clan. The young girl, who was named Princess Wake by the fisherman, is still a source of worship on Gogo Island for many. Gogo Island is where the young girl was first discovered by the fisherman.

The Bible: UFOS, Extraterrestrials, and Gods from the Heavens

The Catholic church has long been said to be hiding knowledge of ancient alien visitation. The Vatican has always looked to the stars, charted the skies and been in possession of some of the most sophisticated telescopes on the planet. They currently own and operate the Vatican Observatory Research Group (VORG) which is a 1.8m Alice P. Lennon Telescope with its Astrophysics Facility, known together as the Vatican Advanced Technology Telescope (VATT), at the Mount Graham International Observatory (MGIO) in southeastern Arizona. The telescope has been "nicknamed" LUCIFER 1 and provides a powerful tool to gain insights into the universe ranging from the Milky Way to extremely distant galaxies. Why does the Vatican feel the need to own and operate an enormous high-powered telescope? Currently, the Vatican owns and operates two high powered telescopes, one at Castel Galfondo, approximately 25 miles south of Rome, and the other "Lucifer" located in Graham County, Arizona. If you look to scripture, the Vatican has had a relationship with other worldly beings since the beginning of recorded time. The following passages from the bible involve encounters with God or "Gods". Could they actually be interrupted as ancient contact with extraterrestrial life?

The Bible: King James Version

Genesis 1:26

26 And God said, "Let us make man in our image, after our likeness: and let them have domination over the fish of the sea, and over the foul of the air, and over the cattle, and over all the earth, and over every creeping thing that creepeth upon the earth.

Genesis 6:1-22

1 And it came to pass, when men began to multiply on the face of the earth, and daughters were born unto them 2 That the sons of God saw the daughters of men that they were fair, and they took them wives of all which they chose. 3 And the LORD said, My spirit shall not always strive with man, for that he also is flesh: yet his days shall be a hundred and twenty years. 4 There were giants in the Earth in those days; and also, after that, when the sons of God came in unto the daughters of men, and they bear children to them, the same became mighty men which were of old, men of renown. 5 And God saw the wickedness of man was great in the earth, and that every imagination of the thoughts of his heart was only evil continually. 6 And it represented the LORD that he had made man on the earth, and it grieved him at his heart. 7 And the LORD said, "I will destroy man whom I have created from the face of the earth: both man and beast and the creeping thing, and the fowls of the air; for it repented me that I have made them. 8 But Noah found grace in the eyes of the LORD. 9 These are the generations of Noah: was a just man, and perfect in his generations, and Noah walked with God. 10 And Noah begat three sons: Shem, Ham and Japheth. 11 The earth also was corrupt before God, and the earth was filled with violence. 12 And God looked upon the earth and behold, it was corrupt; for all flesh had corrupted his way upon earth. 13 And God said to Noah, the end of all flesh is come before me; for the earth is filled with violence through them; and behold, I will destroy then with the earth. 14 Make thee an ark of gopher wood; rooms shalt thou make in the ark, and shalt pitch it within and without

pitch. **15** And this is the fashion which thou shalt make it of: The length of the ark shall be three hundred cubits, the breadth of it fifty cubits and the height of it thirty cubits. **16** A window shalt thou make to the ark, and in a cubit shalt thou finish it above; and the door of the ark shalt thou set in the side thereof; with lower, second, and third stories shalt thou make it. **17** And behold, I, even I, do bring a flood of waters upon the earth to destroy all flesh, wherein is the breath of life, from under heaven; and everything that is in the earth shall die. **18** But with thee will I establish my covenant and thou shalt come into the ark, thou and thy sons and thy wife, and thy son's wives with thee.
19 And of every living thing of all flesh, two of every sort shalt thou bring into the ark, to keep them alive with thee; they shall be male and female. **20** Of fowls after their kind and of cattle after their kind, of every creepy thing of the earth after his kind, two of every sort, shall come unto thee, to keep them alive. **21** And take thou unto thee of all food that is eaten, and thou shalt gather it unto thee; and it shall be food for thee, and for them.
22 Thus did Noah; according to all that God commanded him, so he did.

Genesis 15:17

17 And it came to pass, that, when the sun went down, and it was dark, behold, a smoking furnace, and a burning lamp that passed between those pieces.

Exodus 33:9

9 And it came to pass, as Moses entered into the tabernacle, the cloudy pillar descended and stood at the door of the tabernacle, and the Lord talked with Moses.

Leviticus 24:22

22 Ye shall have one manner of law, as well for the stranger, as for one of your own country: for I am the LORD your God."

II Kings 2:11

11 And it came to pass, as they still went on, and talked, that, behold, there appeared a chariot of fire and horses of fire, and parted them both asunder, and Elijah went up by a whirlwind into heaven.

Nehemiah 9:6

6 Thou, even thou, art LORD alone; thou hast made heaven, the heaven of heavens, with all their hosts, the earth and all things that are therein, the seas and all that is therein, and thou preservest them all; and the host of heaven worshippeth thee.

Psalms 68:17

17 The chariots of God are twenty thousand, even thousands of angels, the Lord is among them; Sinai in the holy place.

Isaiah 13:5

5 They come from a far country, from the ends of the heaven, even the LORD and the weapons of his indignation, to destroy the whole land.

Isaiah 60:8

8 Who are these that fly like a cloud, and as the doves to their windows?

Ezekiel: 1-28

1 Now it came to pass in the thirtieth year, in the fourth month, in the fifth day of the month, as I was among the captives by the river of Chebar, that the heavens were opened, and I saw visions of God. **2** In the fifth of the month, which was the fifth year of the King Jehoiachin's captivity **3** The word of the Lord came expressly unto Ezekiel the priest, the son of Buzi, in the land of the Chaldeans by the river Chebar, and the hand of the LORD was on him. **4** And I looked, and, behold, a windstorm came out of the north, a great cloud and a fire infolding itself, and a brightness was about it, and out of the midst thereof as the colour of amber, out of the mist of the fire.   **5** Also out of the midst there of, came the likeness of four living creatures, and this was their appearance; they had the likeness of a man.   **6** And everyone had four faces, and everyone had four wings.   **7** And their feet were straight feet; and the sole of their feet was like the sole of a calf's foot. A they sparkled like the color of burnished brass. **8** And they had the hands of a man under their wings on their four sides; they four had their faces and their wings.   **9** Their wings were jointed one to another, they turned not when they went; they went everyone straight forward.

10 As for the likeness of their faces, they four had the face of a man, and the face of a lion, on the right side: and they four had the face of an ox on the left side; they four also had the face of an eagle.   **11** Thus were their faces: and their wings were stretched upward; two wings of every one were jointed one to another, and two covered their bodies.   **12** And they went every one went straight forward. Whither the sprit was to go, they went, and they turned not when they went.

13 As for the likeness of the living creatures, their appearance was like burning coals of fire and like the appearance of lamps. It went up and down among the living creatures; and the fire was bright, and out of the fire went forth lightning.   **14** And the living creatures ran and returned as the appearance of a flash of lighting.   **15** Now as I beheld the living creatures, behold one wheel upon the earth by the living creatures, with his four faces.   **16** The appearance of the wheels and their work was like unto the color of a beryl; and they had one likeness: and their appearance and their work was as it were a wheel in the middle of a wheel.   **17** When they went, they went upon their four sides; and they turned not when they went.   **18** As for their rings, they were so high that they were dreadful. And their rings were full of eyes round about them four.

19 And when the living creatures went, the wheels went by them: and when the living creatures were lifted up from the earth, the wheels were lifted up.   **20** Whithersoever the spirit was to go, they went, thither was their spirit to go; and the wheels were lifted up over against them, for the spirit of the living creature was in the wheels.   **21** When those went, these went, and when those stood, these stood, and when those were lifted up from the earth, the wheels were lifted up over against them, for the spirit of the living creatures was in the wheels.   **22** And the likeness of the firmament upon the heads of the living creatures was in the colour of the terrible crystal, stretched forth over their heads above.   **23** And under the firmament were their wings straight, the one toward the other, every one had two, which covered on this side, and every one had two, which covered on that side, their bodies.   **24** And when they went, I heard the noise of their wings, like the noise of great waters, as the voice of the Almighty, the voice of speech, as the noise of an host, when they stood they let down their wings.   **25** And there was a voice from the firmament that was over their heads when they stood and had let down their wings. **26** And above the firmament that was over their heads was the likeness of a throne, as the appearance of a sapphire stone; and upon the likeness of the throne was the likeness as the appearance of a man above upon it.   **27** And I saw as the colour of amber, as the appearance of fire round about within it, from the appearance of his loins even upward, and from the appearance of his loins even downward, I saw as it were the appearance of fire, and it had brightness round about.   **28** As the appearance of the bow that is in the cloud in the day of rain, so was the appearance of the brightness round about. This was the appearance of the likeness of the glory of the

LORD, and when I saw it, I fell upon my face, and I heard a voice of one that spake.

Amos 9:2-3

2 Though they dig into hell, thence shall mine hand take them, though they climb up to heaven thence I will bring them down. **3** And though they hide themselves in the top of Carmel, I will search and take them out thence; And though they be hid from my sight in the bottom of the sea, thence I will command the serpent, and he shall bite them.

Zechariah 5:1-2

Then I turned, and lifted up mine eyes, and looked, and behold, a flying roll. **2** And he said unto me, What sesst thou? And I answered, "I see a flying roll; the length thereof is twenty cubits, and the beneath thereof ten cubits.

John 18:36

36 Jesus answered, "My kingdom is not of this world. If my kingdom were of this world, then would my servants fight, that I should not be delivered to the Jews: but now is my kingdom not from hence.

Acts 19:35

35 And when the townclerk had appeased the people, he said, Ye men of Eph'e-sus, what man is there that knoweth not how that the city of the E-phe'sians is a worshipper of the great goddess Diana, and of the image which fell down from Jupiter?

Revelation 9:7-11

7 And the shapes of the locusts were like unto horses prepared unto battle: and on their heads were as it were crowns like gold, and their faces were as the faces of men.
8 And they had hair as the hair of women, and their teeth were as the teeth of lions.

9 And they had breastplates as it were breastplates of iron; and the sound of their wings was as the sound of chariots of many horses running into battle. **10** And they had tails like unto scorpions, and there were stings in their tails; and their power was to hurt men five months. **11** And they had as king over them, which is the angel of the bottomless pit, whose name in the Hebrew tongue hath his name is A-polly-on (translated is destroyer.)

Revelation 12:1-8

1 AND there appeared a great wonder in heavens; a woman clothed with the sun, and the moon under her feet, and upon her feet a crown of twelve stars. **2** And she being with child cried, travailing in birth, and pained to be delivered. **3** And there appeared another wonder in heaven; and behold a great red dragon, having seven heads and ten horns, and seven crowns upon his heads. **4** And his tail drew the third part of the stars of heaven, and did cast them to the earth: and the dragon stood before the woman which was ready to be delivered, for to devour her child as soon as it was born. **5** And she brought forth a man child, who was to rule all nations with a rod of iron: and her child was caught up unto God, and to his throne. **6** And the woman fled into the wilderness, where she hath a place prepared for God, that they should feed her a thousand two hundred and three score days. **7** And there was war in heaven: Michael and his angels fought against the dragon: and the dragon fought and his angels **8** prevailed not; neither was their place found any more in heaven.

Revelation 12:12

12 Therefore rejoice, ye heavens and ye that dwell in them. Woe to the inhabitants of the earth and of the sea! For the devil is come down unto you, Having great wrath, because he knoweth that he hath but a short time.

Revelation 13:1

And I stood upon the sand of the sea, and saw a beast rise up out of the sea, having seven heads, and ten horns and upon his horns ten crowns, and upon his head the name of blasphemy.

Chapter 2

Known Underwater Human Activity

Humans have been interested in the mysteries of what exists deep underwater, out of one's sight and access for centuries. With 70 percent of the Earth's surface covered with water, most of it is inaccessible without the use of technology. Our oceans, seas, rivers, and lakes hide mysteries and treasures deep below their surfaces, leaving almost 95 percent of this Earth's real estate "still unexplored by mankind" according to NOAA. Underwater access has always been and still remains difficult to inhabit. Some deep vast ocean trenches, even in this age of technology, are still unattainable. Before we explore the possible underwater capabilities and inhabitation of the Earth's waterways by alien races, we must first know and understand what humans were capable of in the past and what our limitations are today. The following are just a few examples of old and new human technology.

Early Submarines

Beginning in ancient times, humans sought to operate under the water. From simple submarines to nuclear underwater powered vessels, we have searched for a means to remain safely underwater to gain the advantage in warfare, resulting in the development of the submarine. The concept of underwater combat has deep roots. There are images of men using hollow sticks to breathe underwater for hunting at temples at Thebes, but the first known military use occurred during the siege of Syracuse (415-413 BC),

where divers cleared obstructions, according to the "History of the Peloponnesian War." At the siege of Tyre (332 BC) Alexander the Great used divers, according to Aristotle. Later legends from Alexandra, Egypt from the 12th century AD, suggested that Alexander conducted reconnaissance using a primitive submersible in the form of a diving bell as depicted in a 16th century Islamic painting.

The first submersible to be actually built-in modern times was constructed in 1605 by Magnus Pegelius. Its fate was to become buried in mud. The first successful submarine in 1620 was propelled by oars and is thought to have incorporated floats with tubes to allow air down to the rowers.

The first air independent and powered submarine was the Ictineo 11, designed by Narcis Monturiol. Originally launched in 1864 as a human-powered vessel, propelled by 16 men, it was converted to peroxide propulsion and steam in 1867. The 46-foot craft was designed for a crew of two, could dive to 98 feet, and demonstrated dives of two hours. The first submarine that did not rely on human power for propulsion was the French Navy submarine Plongeur, launched in 1863, and equipped with a reciprocating engine using compressed air from 23 tanks at 180 psi. It could dive to 33 feet.

The USS Alligator (1861) was the most well-known confederate submarine. It was designed by a French inventor named Brutus de Villeroi. While under tow on the way to Charleston, South Carolina, this submarine was lost in a storm. Submarine development continued on in the Confederacy under the direction of the Secret Service rather than under the Navy. Most Union submarine development was done with the goal of clearing obstructed harbors, while most Confederate submarine development had the more military purpose of breaking up the Union blockades. Source: AmericanCivilWarStory.com

Figure 2 The USS Alligator
Source: Americancivilwarstory.com

The first U.S. nuclear submarine was the Nautilus. It was constructed under the direction of U.S. Navy Captain Hyman G. Rickover, a Russian-born engineer who joined the U.S. atomic program in 1946. In 1947, he was put in charge of the navy's nuclear-propulsion program and began work on an atomic submarine. Rickover succeeded in developing and delivering the world's first nuclear submarine in 1952, the Nautilus, commissioned on September 30, 1954. It first ran under nuclear power on the morning of January 17, 1955.

Today's U.S. Navy

The Four Classes of Navy Submarines: navy.com

"Stealthy, agile and armed with some of the most powerful weapons on the planet, Navy submarines and their crews play a number of roles in both war and peace time: attack, surveillance, commando insertion, research and nuclear deterrence. Submarine operators are known as the "Silent Service" where standards are incredibly high, and victories are often kept secret. Navy submarines are some of the most high-tech vessels in the world. They can insert Seal teams into hostile target areas, launch ballistic missiles, take out enemy subs and ships, perform reconnaissance and rescue missions, and also serve as a platform for nuclear weapons."

Attack Submarines (SSN): Hunters Beneath

The Navy deploys three classes of these sleek subs: the Los Angeles, Seawolf (only 3 commissioned then replaced by the Virginia class), and Virginia. All are capable of performing seek-and-destroy missions on enemy ships and subs, surveillance and reconnaissance, irregular warfare, convert troop insertion, mine, and anti-mine operations and more. Plus, each is armed with Tomahawk cruise missiles to stealthily strike targets from far out.

Ballistic Missile Submarines (SSBN): The Backbone of Strategic Deterrence

"Boomers" may have the most important mission in the Navy: strategic nuclear deterrence. They were built for it as their sole role. Capable of operating underwater for months on end, Ballistic Missile Submarines require crews that can work together under any circumstance. They can defend themselves individually with torpedoes, but what they do for the defense of America cannot be measured.

Guided Missile Submarines (SSGN): Precision from Beneath

To complement the potent guided-missile platform of Cruisers and Destroyers, the Navy needed something a little stealthier. Few things are as stealthy as an Ohio-Class boomer. Guided Missile Submarines were converted from class and can carry over 150 Tomahawk missiles plus transport and support Navy Special Operations forces.

Deep Submarine Rescue Vehicles (DSRV): The Sub Fleet's Life Vest

When trouble happens beneath the waves, there is only one vessel to call on: the DSRV. They perform underwater rescue operations on disabled submarines for the U.S. or foreign navies. They are designed for quick deployment in the event of a submarine accident-transportable by truck, aircraft, ship, or specially configured attack submarine. Many sailors owe their lives to the capabilities of a DSRV and its crew.

The Newest U.S. Nuclear Virginia-Class Submarine

The $2.6 billion USS South Dakota (SSN 790) is the newest, most-advanced addition to the US Navy's Virginia-class fleet of nuclear-powered, fast-attack submarines. The South Dakota was commissioned February 2, 2019, as the Navy's 17th Virginia-class submarine and the seventh Virginia -class Block III submarine. They carry advanced-capability torpedoes and Tomahawk land-attack cruise missiles, and are to be armed with a powerful laser as well. These submarines make their own water, oxygen and sustained fuel source and are the fastest, quietist, sleekest submarine in existence. This Virginia class submarine controls its periscope from a mounted joystick and can launch Navy

Seal team members from below the surface. Estimated top secret speed is 25 knots (approx. 30mph), test depth 1600 feet with a collapse depth being at 2400 feet. The down time is as long as supplies last which is approximately 90 days. The Navy currently has 69 submarines of various classes and hopes to have a total of 66 Virginia-class submarines by 2048. At present France, China, Russia and the US are the only four nations on Earth with the capability to build, arm and launch a submarine.

Underwater Research Facilities

Conshelf

Ever since Jacques Cousteau's 1962 Conshelf I project, humans have been living on the sea floor for purposes of science, innovation, and exploration. In 1963 Conshelf II was launched. It was essentially a small village built on the floor of the Red Sea at 33 feet deep. Scientists could live underwater and do research for a week at a time, record time: 30 days. The third and final Conshelf III mission saw six oceanauts descend to a habitat submerged in over 330 feet of water. Conshelf III was located in the Mediterranean Sea, between Nice and Monaco. The oceanauts spent three weeks there performing industrial tasks on a mock oilrig. Although this mission was deemed successful, it also signaled the end of the Conshelf missions, and of Cousteau's support for marine exploration. Although the missions proved that humans were capable of living and working underwater, further advances in technology meant that these tasks would be carried out in the future by robotic devices.

Aquarius

Currently located in the Florida Keys, there is a high-tech undersea laboratory called Aquarius. This laboratory is also an undersea home for scientists studying the marine environment and is owned and operated by Florida International University (FIU). The underwater habitat currently sits in about 60 feet of water 8 miles offshore of Key Largo, Florida, on a sand patch adjacent to deep coral reefs in the Florida Keys National Marine Sanctuary. On June 1, 2014, in a fitting salute to the revolutionary work of his grandfather Jacques, Fabien Cousteau led a team on a 31-day expedition to Aquarius, during which time they collected enough data for 10 scientific papers. Fabien Cousteau's Aquarius expedition Mission 31 illustrates how times have changed since his grandfather's Conshelf missions. Instead of investigating how to exploit marine resources, this mission was dedicated to studying the effects of climate change and pollution upon the marine life, reefs, and water.

This undersea laboratory is marked by a large buoy floating on the ocean surface with a warning to stay at least 300 feet away. My family and I happened upon it one afternoon after returning from a day of scuba diving. Realizing what it was and what we had found, we respectfully departed from the area excited to realize what was a mere 60 feet below us.

Figure 3 Aquarius laboratory in the Florida Keys
Photo: www.fiuaquarius.net

The Jules

The Jules is an undersea Lodge located in Key Largo, Florida. This is the world's only underwater hotel where scuba diving is the only way to get to your room. This single room hotel is a steel-and-glass facility sitting 30-foot-deep with the entrance at 21 feet. Lodging ranges from 3 hours to overnight and is located off Overseas highway in the Florida Keys just one mile north of John Pennekamp Coral Reef State Park. On October 3, 2014, at 12:08 EST, biology professor Bruce Cantrell and biology instructor Jessica Fain entered the habitat as a classroom project presented by Roane State Community College in Harriman,

Tennessee, and the Marine Resources Development Foundation. This was done to bring awareness of issues effecting the oceans through a unique educational experience by both living in and working from underwater. The professors hosted a weekly program featuring interviews with leading scientists and explorers covering topics such as conservation and undersea exploration. They remained underwater for a record 73 days, 2 hours, and 34 minutes, not surfacing until December 15, 2014, at 13:42 EST.

My family and I had a condo in Key largo for many years and frequently had breakfast at the small restaurant next door to The Jules. The lobby of the Jules had a nice gift shop and live cameras where you could view the underwater hotel and its guests 20 plus feet below (if guest permission was granted).

Tunnels

These are just two examples of underwater transportation tunnels. There are many other tunnels across the world which make their way under bodies of water. Tunnels have been used by the United States, Russia, China, and many other major world powers as access to submarine bases dating back before WWII. Some of these underwater tunnels have miles in length access for underground military bases, secret submarine bases and other secret military facilities including those used to secure high ranking government officials in times of disaster or invasion.

The Holland Tunnel is a highway tunnel under the Hudson River between Manhattan in New York and Jersey City, New Jersey. Its two tubes carry eastbound and westbound Interstate 78. The tunnel opened in 1927 as the first of two vehicular tunnels under the river the other being the Lincoln Tunnel. Both are operated by the Port Authority of New York and New Jersey.

The Channel Tunnel which opened to the public on November 14, 1994, is a 31.35-mile rail tunnel linking Folkestone, Kent in the United Kingdom, with northern France beneath the English Channel at the Strait of Dover. At its lowest point, it is 250 feet deep below the seabed, and 380 feet below sea level. At 37.923.5 miles, the tunnel has the longest undersea portion of any tunnel in the world, although the Seiken Tunnel in Japan is both longest undersea at 53.85 miles and deeper at 790 feet below sea level. The speed limit for trains in the tunnel is 99 mph. Tunnels are built with tunnel boring machines. Known as a "mole" these machines are used to excavate tunnels with a circular cross section through a variety of soil and rock. Modern TBMs (tunnel boring machines) typically consist of the rotating cutting wheel called a cutter head, followed by a main bearing, a thrust system and trailing support mechanisms. The type of machine used depends on the particular geology of the project, the amount of ground water present and other factors. While the use of TMBs relieves the need for large numbers of workers at high pressures, a caisson system is sometimes formed at the cutting head for slurry shield TBMs. Workers entering this space for inspection, maintenance and repair need to be medically cleared as "fit to dive" and trained in the operation.

Two fantastic books by Richard Sauder, Ph.D. titled "Underwater and Underground Bases" and "Hidden in Plain Sight," document Sauder's years of research and findings of the use of underwater bases and facilities and the government secrets behind them.

Chapter 3

Well - Documented Water UFO Sightings

Figure 4 Battle of Los Angeles, LA Times

Figure 5 Battle of Los Angeles, LA Times

The 1942 Battle Los Angeles

(As reported by the press: February 25, 1942, 2:36am)

The sudden appearance of an enormous round object triggered all of LA and most of Southern California into an immediate wartime blackout with thousands of Air Raid Wardens scurrying all over the darkened city while the drama unfolded in the skies above...a drama which would result in the deaths of five people and the raining of shell fragments on homes, streets, and buildings for miles around. The command in San Francisco confirmed and reconfirmed the presence over the Southland of unidentified planes. Relayed by the Southern California sector office in Pasadena, the second statement read: "The blackout in the Los Angeles area for

several hours this a.m. have not been identified." aircraft which caused the Insistence from officials who heard and saw the activity spread countless varying stories of the episode. The spectacular anti-aircraft barrage came after the 14th Interceptor Command ordered the blackout when strange craft were reported over the coastline. Powerful searchlights from countless stations stabbed the sky with brilliant probing fingers while anti-aircraft batteries dotted the heavens with beautiful, if sinister, orange bursts of shrapnel.

The object that triggered the air raid alarm had drawn 1433 rounds of ammunition from the coast artillery to no effect. When it moved at all, the object had proceeded at a leisurely pace over the coastal cities between Santa Monica and Long Beach, taking about 30 minutes of actual flight time to move 20 miles; then it disappeared from view.

Figure 6 Battle of Los Angeles Photo: LA Times

Testimony was given by a female Air Raid Warden; "It was huge! It was just enormous! And it was practically right over my house. I had never seen anything like it in my life!" she said. "It was just hovering there in the sky and hardly moving at all. It was a lovely pale orange and about the most beautiful thing you've ever seen. I could see it perfectly because it was very close. It was big! They sent fighter planes up and I watched them in groups approach it and then turn away. They were shooting at it, but it didn't seem to matter. It was like the Fourth of July but much louder. They were firing like crazy, but they couldn't touch it." "I'll never forget what a magnificent sight it was. Just marvelous. And what a gorgeous color!"

The color may have been gorgeous but five people died that night, three from automobile accidents due to the rush and panic, and two others of heart attacks.
There were also several related injuries reported including a radio announcer who ran into an awning and suffered a gash over one eye, a police officer who cut his right leg while kicking in the window of a lighted Hollywood store, and injuries suffered by several air-raid wardens. One fell from a wall while looking into a lighted apartment and broke a leg, another jumped a 3-foot fence to reach a lighted house and sprained an ankle, and still another fell down his own front stairs and broke an arm.

Structural damage was also widespread. This was caused by antiaircraft shells that failed to explode in air but did so when they struck the ground, demolishing a garage, a patio, and blowing out the tires and causing extensive damage to parked automobiles and several homes and businesses.

The next day the Secretary of the Navy had to face the embarrassment when the publicly stated that there had been no air raid, no enemy planes, and that the whole incident was just a "case of jitters". The army was accused of shooting up an empty sky causing havoc and widespread damage, the sheriff was also embarrassed because he had assisted the FBI in rounding up several Japanese nurserymen and gardeners who were supposedly caught in the act of signaling the enemy invaders.

At the end of WWII an Army document was released explaining the 1942 Los Angeles incident:
1. Numerous weather balloons had been released over the area that night. They carried lights for tracking purposes, and these "lighted balloons" were mistaken for enemy aircraft.
2. Shell bursts illuminated by searchlights were mistaken by ground crews for enemy aircraft.

Final official outcome: Weather Balloon from Long Beach

After the war, the Japanese declared that they had flown no airplanes over Los Angeles on that date.

Maury Island, June 21, 1947

(Reported by Harold Dahl) June 21, 1947, at approximately 2:00pm while on his patrol boat with two men, his son, and their dog, Dahl's boat was approaching the east shore of Maury Island. As Dahl and the others looked in the sky, they saw six objects floating about two thousand feet above their ship. The objects were made of some reflective metal, doughnut shaped, approximately one hundred feet in diameter with the center holes being about twenty-five feet in diameter. Dahl also saw round portholes and what he thought was an observation window. Five of the craft circled over the sixth, which dropped slowly then stopped and hovered about five hundred feet above the water.

Dah, being afraid the center aircraft was going to crash into him, beached his boat. Once ashore, Dahl managed to take several pictures with his camera. The lower ship stayed in position for about five minutes, with the others still circling above. One of the ships left the formation and moved down, touching the lower ships. The two craft kept contact for several minutes, until Dahl said he heard a "thud sound". Without warning, thousands of pieces of what he thought were newspapers began dropping from the inside of the center ship.

Most of the hot debris landed in the bay causing the water to sizzle. Some of the falling debris hit the beach which Dahl managed to recover. He gathered a few pieces, he described as white, lightweight metal. Along with the white metal, the ship dropped what he estimated to be about twenty tons of a dark metal, which he stated, "looked like lava rock". When this lava rock hit the water, its extreme high temperature caused the water to steam. The four of them took cover after several

pieces landed on the boat. Some of the hot debris hit his son on the arm, burning him severely, another piece hit the dog killing it. After this rain of metal, all the craft rose into the air and headed west out to sea together. Dahl went to his boat and tried to radio for help, but his radio did not work. They sailed back toward their dock, in the process, dropping the dog over the side of the boat into the water as a burial at sea. Dahl took his son to the hospital for treatment and then contacted his boss Fred Crisman, to inform him of the incident. Dahl gave his boss Crisman the camera to have the film developed. The photos showed the strange air ships, but the negatives had spots on them. The film may have been damaged by exposure to radiation. Crisman claimed that he did not believe Harold Dahl's story, but nevertheless, went back to Maury Island, where he gathered some of the rock samples. Crisman said that while he was gathering the rocks, one of the airships appeared overhead, as if to be watching him.

Harold Dahl told investigators that the next morning, a man wearing a black suit visited him and suggested they go to breakfast together. Dahl drove his own car, following the stranger's new black Buick to a restaurant. While they ate, the stranger asked no questions; instead, he gave a detailed account of what had happened to Dahl the day before. The man in black warned Dahl that bad things would happen to him and his family if he told anyone about the incident. Dahl and Crisman sent a package to publisher Ray Palmer in Chicago. The package contained a box of metal fragments and statements about the strange happenings on the 21st and 22nd of July.

A few weeks later, Palmer contacted Kenneth Arnold who had begun investigating UFOs. Arnold arrived in Tacoma in late July with airline pilot E.J. Smith. The two of them met with Dahl and Crisman, examined Dahl's boat,

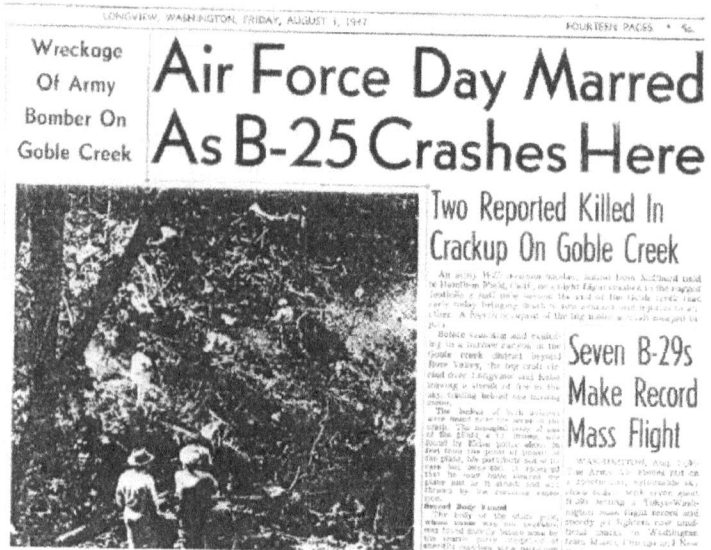

Wreckage Of Army Bomber On Goble Creek

Air Force Day Marred As B-25 Crashes Here

Two Reported Killed In Crackup On Goble Creek

Seven B-29s Make Record Mass Flight

Figure 7 Maury Daily News

and conducted an investigation. On the afternoon of July 31,1947, Captain Lee Davidson and First Lieutenant Frank Brown of the U.S. Army Air Force flew up to Tacoma from Hamilton Field, California. In addition to being pilots, the two men were intelligence specialists. They met with Arnold, Smith, and Crisman for several hours. One of the officers said that he thought there might have been "something" to the story, but they had to leave around midnight. The two officers flew out of McChord Airfield around two o'clock in the morning on a B-25 bomber, with a crew of two other men. About twenty minutes later, the airplane crashed near Centralia, Washington. The two enlisted men managed to parachute to safety, but Davidson and Brown were killed.

Dahl and Crisman said that the AF officers took some of the strange metal onboard. People thought they heard anti-aircraft guns shoot the plane down. The local newspapers and FBI received phone calls stating that the plane was shot down to cover up the information Brown and Davidson had found. Because of the loss of life, the Air Force broadened its investigation,

and the FBI launched their own. The Air Force investigators determined that the crash had been a terrible accident. One of the engines caught fire and the men began bailing out. Before Brown and Davidson could jump out, a wing broke and struck the tail section, which also broke off. The plane went into a spin, trapping the men inside.

Another Air Force investigator spoke with Dahl and Crisman and visited their boat. He stated that the damage he saw did not match the damage the two sailors described. There were no piles of metal on Maury Island, and the existing samples looked like slag from a metal smelter. His conclusion matched that of the FBI investigator: that Dahl and Crisman had faked the incident to gain publicity for a magazine article. The FBI warned Dahl and Crisman that their hoax had not succeeded and that if they dropped the matter, the government would not prosecute the two men for the fraud, which had resulted in the deaths of the two officers. At first Dahl and Crisman went along. They made statements that the story was a fake and simply refused to give interviews on the matter. But a few years later in the January 1950 issue of Fate magazine, Crisman stated that the incident had happened. Kenneth Arnold also included the Maury Island incident in his 1952 book "The Coming of the Saucers."

Today, most people believe that Crisman and Dahl faked the incident, perpetuating a hoax that got out of control. Other people believe that the U.S. Government was behind a conspiracy that may have involved anything from UFOs to dumping nuclear waste in Puget Sound. They believe a shadow government agency sabotaged the B-25 bomber in order to eliminate the investigators and blame Dahl and Crisman.

Catalina Island, California

Catalina Island is twenty miles off the mainland. It is twenty-one miles long, eight miles at its widest point, with a fifty-four-mile perimeter. Catalina's highest point Mount Orizaba sticks 2,069 feet out of the Pacific Ocean which reaches depths of 3,000 feet between the island and the mainland.

July 7, 1947

The first California underwater UFO sighting was reported on July 7, 1947, when two San Raphael teen-agers saw a "flat glistening object" emerged from the water, fly around and then dive back into the water 400 yards from shore.

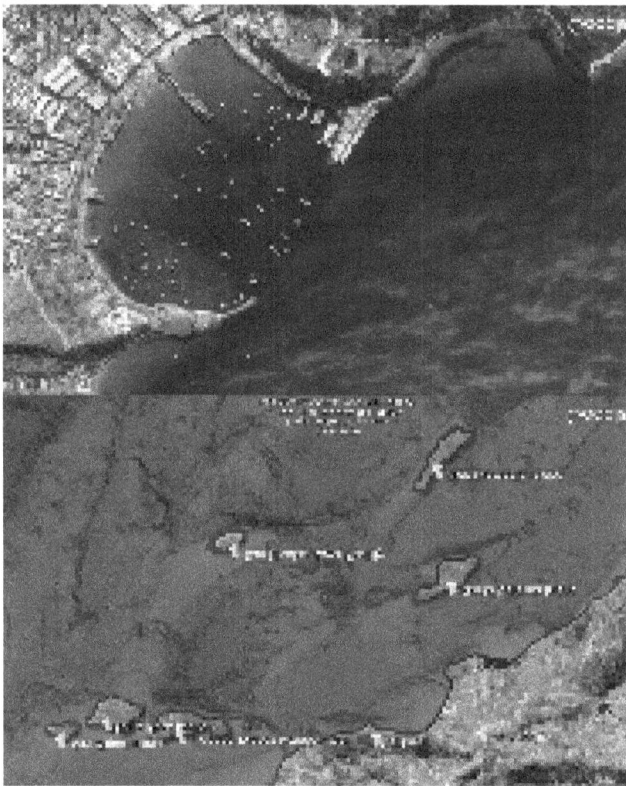

Figure 8 Catalina Island. Photos: Google Earth

July 8, 1947

On July 8, 1947, the very same day that Roswell's Daily Record was reporting the initial "flying disc" story, a remarkable incident reportedly occurred in the skies above Avalon. An article on the front page of the week's issue of the Catalina Islander details an alleged sighting by three visiting Army veterans of six "flying discs" traveling at high speed from the northeast and passing directly over Avalon before disappearing over East Peak. According to the story, the six discs appeared at about 1:00 p.m. and flew in a formation of two sets of three and were witnessed not only by the veterans, but by "hundreds" of others as well. Russo, one of the reported witnesses and an Army Air Corps veteran who had flown 35 bombing missions over Germany with the Eighth Air Force, estimated the velocity of the discs at "850 miles an hour," according to the story. Throughout this same year, numerous steamers reported a mysterious 'undersea mountain' or a 'large mass underwater' which kept appearing and disappearing in various locations in the San Francisco Bay and down the coast.

August 1947

In August of 1947, the Coast Guard received reports of a "strange flaming object," which fell into the sea.

Fall 1947

In the early fall of 1947, steamers going into and out of San Francisco Bay encountered an "undersea mountain" that appeared and disappeared in various locations in the bay. The mysterious mass was reported by several ships many referring to it as a "reef" or "submarine mountain" that appeared overnight. The same mass was reported by another ship, which described what they saw as "a large mass under water, off the Golden Gate." Soon after these many accounts, the mass mysteriously disappeared.

Following the 1947 Incidents Sightings came regularly, in 1951, 1954, 1955, 1957, 1962, 1964, 1970, 1980, 1990, 1991, 1993, 2004 the list goes on, most from the Santa Catalina channel. The sightings involve many highly credible witnesses including lifeguards, security guards, law enforcement officials, military officers, and countless citizens.

November 21, 1951

Harold Wilkins, UFO researcher, investigated a report from November 21, 1951, of several witnesses observing "an unidentified burning object" descending into the ocean off the coast of California.

November 1953

Engineer Frederick Hehr and several others were on Santa Monica beach when they observe a "squadron of saucers" performing maneuvers in the daylight sky over the bay. Later that day, the objects return and perform more maneuvers for a period of about ten minutes.

August 8, 1954

Japanese steamship The Aliki August 8, 1954: Long Beach coastline, several crew members observed an underwater UFO. (Intercepted radio message from the ship) "Saw fireball move in and out of sea without being doused. Left wake of white smoke; course erratic; vanished from sight."

July 10, 1955

On July 10, 1955, at approximately 11:00 a.m., several fishermen off the coast of Newport Beach observe a bluish-silver, cigar-shaped object flying overhead at a "moderate speed and medium altitude." Two and a half hours later, a Washington family of three were sailing thirteen miles off the coast of Newport Beach on their way to Catalina Island when they observe a "perfectly round, gray-white" craft about 2,500 feet above their boat. The craft hovered over their boat, frightened, they radio the Coast Guard, which sent out a plane.

Unfortunately, the craft sped away before the Coast Guard plane arrived.

1955

Santa Monica 1955: Santa Maria residents report a "long silvery object" emerging from the ocean and taking off into space at a high rate of speed.

January 15, 1956

Dozens of residents from Redondo Beach filed reports to police during the evening hours on January 15, 1956, of "a large, glowing object glide down out of the sky and float on the surface of the ocean about 75 yards offshore." Included with the witnesses were a local night watchman, Redondo Beach lifeguards, and police officers from Hermosa Beach. As the multiple witnesses watched, the water around the object starts to "froth" and the UFO sinks beneath the water surface. The glow of the object was so bright, that it could still be seen below the water surface. The Hermosa Beach police officers radioed for assistance. Scuba divers showed up at the scene to investigate, but unfortunately, the object is gone. The area was tested for radiation by one of the officers who was using a Geiger counter, but no radiation registered.

February 9, 1956

Leonard Stringfield obtained an official military report on an incident from February 9, 1956, which states: "Fireball hits water. Submerges." A year prior on the same date, witnesses had reported seeing a fireball descending into the ocean off the coast of Redondo Beach.

November 6, 1957

On November 6, 1957, during the early morning hours in Playa Del Rey, three cars driving along the Pacific Coast Highway suddenly stalled out when a large "egg-shaped object" surrounded in a "blue haze" landed on the beach only a few yards away from PCH.

The three witnesses Richard Kehoe, Ronald Burke, and Joe Thomas got out of their stalled cars to observe the craft. As they did, they saw two "strange-looking men" exiting from the craft. The "beings" were described by the men as having "yellowish-green skin" and wearing "black leather pants, white belts and light-colored jerseys." The "beings" walked up to the witnesses and begin asking them questions. The men were unable to understand the language the "beings" were speaking. After a few minutes, the "beings" returned to their craft and left at a high rate of speed. Later that day at approximately 3:50 p.m., an officer and twelve airmen from Long Beach, observe six saucer-shaped crafts moving across the sky. Two hours later, officers at Los Alamitos Naval Air Station reported seeing "numerous" craft crisscrossing the sky. Long Beach police received more than 100 calls from residents reporting UFOs.

December 1957
While off the coast of San Pedro, in December of 1957, the crew of the British steamship Ramsey observed a large metallic gray disk with antenna-like projections. A photo of the object was taken by one of the crewmen before the craft flew out of range, but when the photo was developed, it was blurry.

1960
As he sat on the rooftop of his Beverly Hills home one evening in 1960, screen actor Chad Everett and two friends observed a lighted object moving back and forth at high speeds over the Pacific Ocean. The witnesses were convinced they saw a UFO because of the quick, right angles moves the craft made.

July 28, 1962
As documented by military personnel; July 28, 1962: The captain of a chartered fishing boat sees lights floating in a stationary position in the water about six miles south of Catalina Island. As the boat captain got closer, the craft was not a Russian submarine as first thought, but as described by the boat captain; "It appeared to be the stern of a submarine," he says. "We could see five men, two in white garb, two in dark trousers and white shirts, and one in a sky-blue jump suit. We passed abeam at about a quarter mile, and I was certain it was a submarine low in the water, steel gray, no markings, decks almost awash, with only its tail and an odd aft-structure showing." Suddenly, the strange craft headed directly towards the fishing boat, seemingly, to ram it. The captain quickly made an emergency turn as the craft pasted them closely at a high rate of speed. The craft emitted no sound, no wake, just a "good-sized swell." The captain promptly contacted the Navy Base which could not identify the craft just witnessed by the boat captain and crew. The sighting was investigated by UFO researchers Coral and Jim Lorenzen who speculated that it may have been a UFO and not a submarine: "The high speed, lack of wake and sound, and the huge swell make this object suspect." One might also mention the odd shape of the submarine itself, its lack of fear of observation, and its aggressive maneuvering."

February 5, 1964
Eleven passengers are rescued by the Coast Guard from their emergency raft following the unexplained sinking of their yacht, the Hattie D. on February 5, 1964. The crew was sailing south down the coast of California from Seattle, Washington, when their yacht either struck or was rammed by an unidentified "metal object." Crewman Carl Jansen says, "I don't care how deep it was...what holed us was steel, and a long piece. There was no give at all."

December 2, 1965
On December 2, 1965, Mrs. Irwin Cohen and her son observe a "glowing red object

descending into the sea off San Pedro" which set off a large cloud of steam as it descended. The Cohens took a few pictures. (Pictures not available)

April 15, 1966

Figure 9 UFO Captured at Cataline Island Photo by Leland Hansen

Leland Hansen, a Navy news crew photographer, while on assignment on April 15, 1966, was able to capture an unidentified craft on video. Hanson was filming Catalina Island from a helicopter, when he noticed something unusual moving across the island over the mountain tops. Under close analysis it was determined that the craft was oval-shaped, with no wings or tail structure, and had no identifiable features. Video analysis determined the craft to be approximately 30-feet in diameter and traveling at an estimated speed of 130-170 miles per hour.

October 1968

In October of 1968, George Hiner was fishing in his boat off the eastern end of Catalina Island. Hiner noticed a strange object and grabbed his binoculars. What he saw he describes as being; a "white-domed shaped object" While watching, the object rises ten feet above the surface of the water, descends and then rises again. Through his binoculars, Hiner can see a strange parachute-like device beneath the object, which gently descends and then sinks beneath the waves.

1970

MUFON investigator Bill Hamilton, investigated an anonymous report from 1970. A gentleman sailing from Catalina Island to San Pedro Harbor observed a "metallic saucer' with four "hemispherical pods" underneath it passing just a few hundred feet above his boat at he sailed.

1973

In May 1973 an art director while driving along the Pacific Coast Highway in Santa Monica just before dawn, reported seeing a craft shining down a beam of light. "The UFO was over where the beach, it was hovering I would say maybe a hundred, two hundred feet in the air. It was silver. It was your basic UFO. It was definitely completely metallic with a silver dome on top and a silver dome on the bottom of it, like two plates put together. And it had little lights around it." Additional witnesses were able to confirm the sighting.

January 26, 1980

A strange unknown craft off Rippers Cove, Catalina Island was the cause of an airplane crash on January 26, 1980. Pilot Noah Felice, claimed that a UFO on top of the water, shot a "beam of yellow light" at his small plane. The UFO then shot a second beam of light hitting his plane which Felice claimed caused the plane's equipment to malfunction. Shortly after, they crashed into the Pacific Ocean resulting in severe injuries to himself, other passengers, and the death of his cousin Mark. Felice was flying a single-engine private plane when it crashed in 60 feet of water off Catalina Island on a Saturday, killing a visiting New Yorker and critically injuring his Van Nuys cousin Mark. The third person, whose identity was not immediately known, was reported by the Coast Guard as "missing". The dead man was identified as Mark Anthony Felice, 18. He was visiting his cousin, Noah Felice, 28. Nearby skindivers pulled Noah from the water unconscious but managed to revive him. According to the Coast Guard, his cousin Mark had apparently died immediately. Noah Felice was taken by helicopter to Torrance Memorial Hospital, where he was under intensive care with several broken bones and other injuries.

June 1980

Therapist Linda Susan Young and a friend are driving along the PCH (Pacific Coast Highway) in Santa Monica one evening in June of 1980 when Young observes an unusually bright light hovering several miles out above the Pacific Ocean. Young was confused by the object's appearance asking her friend; "I said to the guy with me, what do you suppose this is?' And he turned around and looked at it. And he only saw it for a second when it just shot straight up in the air and blinked out. It didn't look like it went far enough to disappear from view, like a distance. It just sord of stopped. It just stopped being there. I have always assumed it was a UFO."

1980s

 Anonymous witness described as a senior electronics engineer: The witness stated that he was sailing on a foggy day between Santa Barbara Island and Santa Cruz Island sometime in the 1980's, when he observed a "fluorescent green colored light" ahead of him in the mist. Believing it was another ship using bright lights to navigate the fog, the witness stopped, and waited for the vessel to pass. As the vessel approached his position, he is still unable to distinguish what type of vessel he was looking at. As the vessel approached, still being a quarter of a mile away heading directly toward him, the witness realized as stated; "I finally realized that this dumb thing was underwater, I'm guessing it was, I don't know, maybe 300 feet in diameter, but I couldn't get any vertical dimension on it because it was under me in

the water. It literally passed directly underneath me." Sailing a fully equipped, 38-foot sailboat, the witnessed watched in awe as the vessel passed beneath him. The witness stated that he took several readings from the depth sounder, determining that the object was approximately 100 feet deep, then without warning, both boat depth finders quit functioning. The witness checked his compasses stating, "All three of them were slowly rotating and I wasn't. I tried calling the Coast Guard and the radio was dead." The vessel continued its path, moving away and disappearing. The witness badly shaken from the incident, checked his equipment which revealed that all the compass mountings were broken. "It was weird. I was just too damned petrified to move."

June 15, 1982

MUFON CMS #72642

While sailing with parents to Catalina Island from Newport Beach, sailing a 44' Gulfstar boat at approximately 8:00pm, the witness and his mother saw a huge yellow object moving in the water under his boat at 50 feet below them moving around 20 knots. The boat was rocking back and forth. The witness, his stepfather and mother were sailing to Catalina Island on a trip at night to the island. Around 8:00 pm his mother yells to his stepdad to look over the port side and asked what is that? The witness looked and saw a huge lighted object moving fast under the water. His mother was very scared and went below. His stepfather was asking the witness what that was? The witness was in the Navy for years and sailed all over the world, so he knows ships and submarines. The incident lasted approximately five minutes with the object just moving from one side then the next in a crisscross pattern. The witness sailed west moving away from it as it moved east. The witness never told anyone about this, nor did he talk about it. His stepbrother is in the Navy and worked on a LA class submarine. The witness spoke of the incident one night to his stepbrother and was told to "just forget it and to not go there". The object was 50 foot in a circle and bright as more watts than the witness had ever seen. The witness is an electrician and works on lighting and described the object as very bright and giving off heat. The witness's parents stopped sailing after that summer.

Summer 1988

Professional photographer Kim Carlsberg witnessed a strange object in the summer of 1988. While outside viewing the ocean from her Malibu beachfront home, she observed a "darting, star-like object" which quickly moved directly toward her. Carlsberg stated, "The brilliant point of light advanced until it became a luminous sphere some fifty feet in diameter. It ominously hung in the air a hundred feet from my window, the apparent standoff lasted no more than a minute before the sphere departed as quickly as it appeared. It tore away diagonally through the night sky and vanished."

Late 1989

Beginning in late 1989 and investigated by MUFON Investigator Bill Hamilton, numerous witnesses in Marina Del Rey were reporting repeated encounters with "strange blue-green lights in the water." Hamilton stated, "In 1989 and again in 1990, witnesses have seen as many as twenty events an hour. One large light appeared to be as much as 100 feet in diameter. This large light spawned baby lights no larger than 10 to 12 feet in length. These lights were seen to move swiftly under the ocean's surface some 500 to 1000 feet from the coastline in Abalone Cove. One of the lights was reported to have emerged from the water."

Summer 1990

In the summer of 1990, on a late afternoon, Toshi Inouye, a private pilot along with his student, observed a large, red, glowing cigar-shaped object which was hovering near their plane as they fly over the Santa Monica Bay. Inouye stated, "It was standing still in the air, glowing red. We were kind of stunned, we didn't know what to do." Before Inouye could contact the airport control tower, the craft quicky flew away out of sight.

May 1990

In May of 1990 off the coast of Malibu, California, a sighting report was filed with MUFON of a strange craft seen in the direction of Catalina Island 48 miles west across the Pacific Ocean. The witness stated, "I was surfing, with another guy, at County Line (across from Neptune's Net), when it came out of the fog bank just off of shore. The stranger and I were focused on the horizon, awaiting another wave. We didn't know each other, but it's wise to buddy-up when you're in the drink, so there we were, silent, sitting on boards, acknowledging the expectation that if either one would have a problem, the other would be close by. It was a good morning of waves, coming in sets about 7-10 minutes apart. Between the sets, at exactly 12:20 PDT, what looked to me like a brushed aluminum saucer with a bump in the middle (both top and bottom alike), approached the shoreline from out of the fog bank sitting about a mile offshore. It was the size of a US nickel held at arm's length and stopped its lateral approach to the shoreline just inside the fog bank, where it was clear and sunny. I couldn't discern any kind of external rotation, but looking back in my mind, I believe perhaps the 'bumps' were rotating, although I couldn't say why, maybe they had black slits or something (which would be subtle when moving, like a zoetrope or something). It came into the sun, settles (at about 300 feet over the ocean surface), and begins an oscillating descent, much like a flat rock in water, or a leaf towards terra, side to side it drops, for about 3 times. On the 4th, it slowed, tilted towards the shore (us), tilted away, and then, in a blink of an eye, took off at a 60 degree climb into and through the fog bank, out to sea. This happened in the space of about 10 seconds."

Spring 1991

One early morning in the spring of 1991, a man while looking out of the window of his Malibu beachfront home, observed a very bright object floating on the ocean's surface approximately two miles out. The man stated, "It looked like a big prism, kind of various colors out there. I got a telescope out and looked at it." The light was visible for several hours, then just blinked out. In January of 1993, the man reported that the object had returned. "I got the telescope out and looked at it, and it was the same kind of thing, the colors seemed so pure for lack of a better word. They seemed really coherent." The man called the Coast Guard hoping for answers but was given no information on his strange double sighting.

September 19, 1991

MUFON CMS #2790
Four friends went fishing on a 35 ft. private boat out of L.A. Harbor. They left the night of September 17, 1991 and fished all day the 18th. The friends pulled into Avalon for gas then fished off the coast of Avalon. On the night of the 18th, they started back toward L.A Harbor trolling on the night of the new moon on a calm ocean with visibility unlimited. The fishermen witnessed UFO activity all night including what the Space Shuttle captured. They watched one particular UFO for hours that changed shape from a ball of whirling Christmas garland into a defined ship in a vertical position. It slowly rotated 90 degrees and when it appeared to be horizontal to the

horizon it shot across the sky. Released NASA footage shows the object appearing to avoid an attempt to shoot at it or intercept it by suddenly changing direction. The witness claims that he and his friends had been viewing the object before the footage taken and claim that not on film is where the object went. Witness statement: "We had already observed that UFO for at least 2-3 hours prior to the film. Front row seats. It flew like a fly trapped in a cube. It had no definition other than it looked like a glowing ball with Christmas garland racing around the outside. When the UFO changed direction to avoid the intercept it disappeared momentarily. It then reappeared in the same area it had been in earlier and as the glowing ball of garland. One minute sitting in one area of the sky the next minute making incredible maneuvers like a fly trapped in a cube". Their first sighting was at approximately 1030 pm over LAX. But as the night went on, more UFOs appeared and formed groups of 4-6 in the sky which the witness states were also on the released Nasa film.

May 5, 1992

While walking along Malibu Beach on May 5, 1992, two friends observed what they described as a "sort of light-fireball" which they stated, descend from the sky and into the ocean. "It was going at an incredible speed, and it was less than a mile away. It looked like it hit the ocean. Once the object made its way to the ocean's surface, it disappeared, so my guess is that it went underwater."

June 14, 1992

On the day of June 14, 1992, there were reportedly hundreds of UFO's observed rising from the waters off Catalina Island. Independent witnesses claim that there were in excess of two hundred UFO's that emerged from the water, hovering momentarily before accelerating off into the sky at a high rate of speed, all moving overhead in complete silence. There were multiple witnesses, several of whom filed reports with their local police offices, some as far away as Malibu, California which lies 48 miles north. The reports were quickly forwarded to the US Coast Guard who refused to acknowledge any search of the area by their department and dismissed numerous requests to carry on a search of the area of the multiple sightings.

1994

While walking near the coast of Rancho Palos Verdes one night in 1994, two men saw several "glowing disks" floating on the water surface. A few days later, one of the witnesses returned to the same area and saw the disks again. He stated that he also saw several black helicopters in the area at the same time the "glowing disks" were visible on the surface of the water. He was later confronted by unnamed individuals who sternly inform that this area, off Abalone Cove, is "off-limits".

1998

In 1998 a pool cleaner and a military private named were driving along PCH (Pacific Coast Highway) in Malibu. Suddenly, they saw "six black, diamond-shaped objects" moving at a high rate of speed traveling up and down the coast. The two men tried to chase down the strange craft driving quickly down PCH for the next hour hoping to catch another glimpse. One man stated, "We did come across a couple of people who were just sitting in their lawn chairs along the road. I don't know if this has anything to do with it, but they were just sitting there along the side of the road, just looking up." The men didn't stop to ask what the people were looking at, and unfortunately the two did not see these craft again.

May 23, 1998
MUFON CMS #58357

The witness observed a bright glowing orb moving low over the water of the Pacific Ocean, disappearing off the east end of Catalina Island. On May 23, 1989, the witness's wife and two friends were celebrating an anniversary and were just returning to their hotel the Zane Grey, which is located high on the hillside overlooking Avalon and the pavilion. There are a number of open terraces, so the witness was standing on one higher than his wife and friends. They were looking up at him and he was looking at the view of the city. From the right, the west, the ocean side, the witness saw a very bright (brighter than a full moon) circle, orb, about the size of a silver dollar held at arm's length, moving toward the coast of Long Beach, CA. The witness remarked that the object seemed to be in the channel between Catalina and the main land beach, closer to the island. His first thought was if that thing doesn't stop it will crash into the city of Long Beach. The object was very low for the entire time it was in sight, moving in what seemed to be a straight level line over the dark ocean. The witness stated, "I believe it went into the ocean, somewhere off the southeast point of Catalina Island or vanished behind rises in the land at that end of the island. My companions did not see it, they were looking up at me and talking and by the time they turned to look it was gone." The witness is a retired photographer, with 30 years in the aerospace industry. He is very familiar and has photographed every aircraft imaginable and some not so imaginable, and stated, "I have never seen anything like that before or after".

June 12, 1999
MUFON CMS #30495

The witnesses were camping on an isolated beach on an island 26 miles off the coast of California in June of 1999, when they first saw one unidentifiable light and then two shooting around defining physics. Then later, a very large unidentified object about the size of a football field, came down. The witnesses were too frightened to discuss the incident further.

January 11, 2002

An anonymous man: While camping along the coast at Point Mugu, on January 11, 2002, the witness sees a light moving back and forth 100 feet above the water, and two other lights beneath the surface of the sea. The airborne object quickly moves back and forth in tandem for approximately 30 minutes seemingly giving search for something in the waters below. Suddenly, the light in the sky accelerates up and away out of sight, and the two lights in the water below dive down deep disappearing.

January 3, 2004

While standing along the Santa Monica coastline observing the ocean one afternoon on January 3, 2004, a witness observed a "metallic, saucer-shaped craft hovering only a few thousand feet directly above a small yacht which was less than a mile out at sea." The witness was able to take one picture before the craft moved out of sight.

February 22, 2012

MUFON CMS #36012

At approximately 8:00 p.m., the witness was at home and outside when he noticed a very bright object out in the sky over the Pacific Ocean near Santa Catalina Island. "It stood out much brighter than any other star and was varying in its intensity of light like a flame would". At first, the witness thought it might be the ISS (International Space Station) it did not match the description, and the object was much too bright to be any star or planet. He continued to watch the object over the next 45 minutes or so, which remained stationary for the entire time. The witness knew he was not looking at a plane or helicopter and the object did not have any blinking lights. About 20 minutes into the sighting, the object began to

diminish in brightness and disappear completely into some type of vapor cloud in the moonlit sky where it had been. It then reappeared intensifying in brightness until it became very bright again and then eventually disappeared completely. The witness stated, "Just as it had disappeared for good, all of I sudden I see several miles across to the left of where it had been, a very large "arc" of light appeared like a giant flame had just erupted and then went out completely to be replaced by a blinking object which then started to travel right in the direction of where the first object had appeared. It moved slowly until it went past the first objects position completely and onward". This second object did not appear to be a plane or helicopter, but the witness did see several helicopters or planes over the Pacific Ocean in the area while this incident was occurring. The witness stated, "I don't know if any of them were just passing through or were actually "following" this object(s). I have seen these "flaming stars" in the past looking westward out over the Pacific Ocean before and cannot explain what they are".

August 14, 2015, 21:00

MUFON CMS #113092

MUFON Investigator Report

The witness, his wife, and his brother-in-law were camping at Parson's Landing, on the northern tip of Catalina Island during the week of August 10th-14th 2015. While sitting on a bench and looking out at the ocean (the mainland was merely an orange yellow glow in the distance), the witness noticed a golden ball of light making lateral and verticals 'cross-like' movements 60 degrees in the air, approximately 10 miles out at sea (at the mid-point between Catalina Island and the mainland). The lights then split into two lights, golden-white in color, the same magnitude of light as before (magnitude +5 in astronomical terms: 5X the brightness of the brightest star).

The two golden lights then moved together laterally and vertically, in tandem. The two objects then met together again, fused into one object, which then shot down into the sea. When it shot down into the sea, the witness states you could still see the light, but now more diffuse, as though underwater. The light then shot out of the ocean again, hovering brightly in the air, then shot back down, and disappeared into the sea. The witness is a commercial pilot, states he's seen multiple drones, helicopters, and all manner of aircraft, and this was like nothing he's ever encountered before.

Shag Harbor, Canada October 4, 1967

On October 4, 1967, at approximately 11:20 p.m., it was reported that something had crashed into the waters of the Gulf of Maine near Shag Harbor. At least eleven people saw a low-flying lit object head towards the harbor. Multiple witnesses reported hearing a whistling sound, then a "whoosh," and then what sounded like a loud bang or crash. The initial report was made by local resident Laurie Wickens and four of his friends who were driving through Shag Harbor on Highway 3. Suddenly they spotted a large object descending into the waters off the harbor. They quickly decided to move to a better vantage point to see what had occurred. Wickens and his friends saw what they first thought was an airplane floating 250 m (820 ft.) to 300 m (980 ft.) offshore in the Gulf of Maine. Wickens contacted the RCMP (Royal Canadian Mounted Police) detachment in Barrington Passage and reported he had seen a large airplane or small airliner crash into the Gulf of Maine.

The Coast Guard Cutter #101 and many local boats rushed to the spot of the sighting, but by the time they arrived, the light was gone. However, the crewmen of the cutter could still see a yellow foam floating on the surface of the water, indicating to them that something had possibly submerged. Nothing else could be found that night by the Coast Guard, locals or RCMP, so the search was called off at 3:00 a.m. Within 15 minutes of the original emergency call, 10 RCMP officers arrived at the scene. Concerned for survivors, the RCMP detachment contacted the Rescue Coordination Centre (RCC) in Halifax to advise them of the situation, and to ask if any aircraft were reported missing. The object in the gulf was only on the surface for a few minutes so any attempt at rescue or craft identification was not possible, it quickly sank and disappeared from view. Within half an hour of the crash a rescue mission had been quickly assembled by local fishing boats. They went out to the crash site in the waters of the Gulf of Maine off Shag Harbor to look for survivors. No survivors, bodies or debris were found by the fishermen or the Canadian Coast Guard search and rescue cutter, which had arrived about an hour later from nearby Clark's Harbor.

The next morning, RCC Halifax had determined that no aircraft were missing, but were still obligated to continue with the search. Something had definitely crashed into the Gulf of Maine. The captain of the Canadian Coast Guard cutter received a radio message from RCC Halifax that all commercial, private and military aircraft were accounted for along the eastern seaboard, in both Atlantic Canada and New England, so the question for them to ponder was "what would they be looking for." The same morning, RCC Halifax sent a priority telex to the "Air Desk" at Royal Canadian Air Force headquarters in Ottawa, which handled all civilian and military UFO sightings, informing them of the crash and that all conventional explanations such as aircraft, flares, etc. had been dismissed. This was labeled a "UFO Report," unidentified flying

ROYAL CANADIAN MOUNTED POLICE - GENDARMERIE ROYALE DU CANADA

OTHER FILE REFERENCES: REF. AUTRES DOSSIERS:	DIVISION "H"	DATE 7 OCT 67	RCMP FILE REFERENCES: REF. DOSSIERS GRC.
	SUB-DIVISION - SOUS-DIVISION HALIFAX		67-400-23-X
	DETACHMENT - DÉTACHEMENT LUNENBURG		

RE: OBJET:

Unidentified Flying Object.
Sighting of -
Sambro Light, N.S.
(4 OCT 67)

1.   On this date a request was received from the Halifax Sub-Division Section N.C.O., via XJB 84, to contact Capt. Leo Howard MERSEY, of the M/V "J.B. NICKERSON", relative to his sighting of a flying object off Sambro Light on 4 OCT 67. It was further requested that the results of enquiries be forwarded to Barrington Passage Det., in view of a similar sighting in that area.

2.   Capt. MERSEY was interviewed and the following statement obtained:

STATEMENT OF CAPT. LEO HOWARD MERSEY (B: 12 JUNE 22),   Centre, N.S.
Centre, Lun. Co., N.S.   7 OCT 67.

            At about 9 P.M., on the 4 OCT 67, I noticed an object with three flashing red lights. Radar indicated this object to be sixteen (16) miles away. It was very clear that night and we could see the lights of Halifax. At the time our boat was 32 miles south of the Sambro Light and the object was approximately 16 miles north east of us. I would say the object was 16 miles south east of Sambro. At times the Navy do a lot of practising in the area. At the same time there were three other objects on the radar and about 6 miles from the first object. I would say it disappeared about 11:00 P.M., when it went up in the air. I could not see any shape or form to it because of the distance. When it went into the air it only had one flashing light. While the object was on the water, or close to the water, it had three real bright flashing red lights. All the lights on it were red. I could not see any lights on the other three objects as they were only appearing on the radar. It is not unusual to see the Navy, or aircraft, dropping things into the water there. I had never seen anything like that before but it sounds like the thing they are looking for down off Shelburne or Barrington Passage. When the object left it went straight up in the air with only one red light.

Witnessed: D.J. RAHN, 2/Cst.   Signed: Capt. Leo H. MERSEY.

3.   Capt. MERSEY is considered to be a reliable type individual and bears a good reputation in his community.

4.   Barrington Passage Detachment were advised of the foregoing via telephone. A copy of this report is being forwarded direct to that point for their information.

    CONCLUDED HERE.

                                            Cpl.
                                (J.T. Kovacs) #18905.
                                I/C Lunenburg Detachment.

Fleet Diving Unit Atlantic was assembled who, for the next three days, combed the seafloor of the Gulf of Maine off Shag Harbor looking for the craft. Their final report stated that no trace of an object or craft was found.

Captain Leo Howard Mersey

"At about 9:00 p.m. on 4 Oct. 67, I noticed an object with 3 flashing red lights. Radar indicated this object to be 16 miles away. At the time our boat was 32 miles south of the Sambro light and the object was 16 miles north of us. At the same time there were 3 other objects on radar 6 miles from the first object, but they had no lights. When the object disappeared at about 11:00 p.m., it went straight up into the air with only 1 red light visible. I had never seen anything like that before."

object because indeed what crashed was unidentified. The head of the Air Desk sent another priority telex to the Royal Canadian Navy headquarters concerning the "UFO Report" and recommended an underwater search be mounted for possible recovery. The RCN in turn, sent another priority telex assigning the job of search and recovery to Fleet Diving Unit Atlantic. Two days after the incident, a detachment of RCN divers from

Statements by the Divers and Other Witnesses to the Events

"The object that dove into the waters of the harbor had soon left the Shag area, traveling underwater for about 25 miles to a place called Government Point, which was near a submarine detection base. The object was

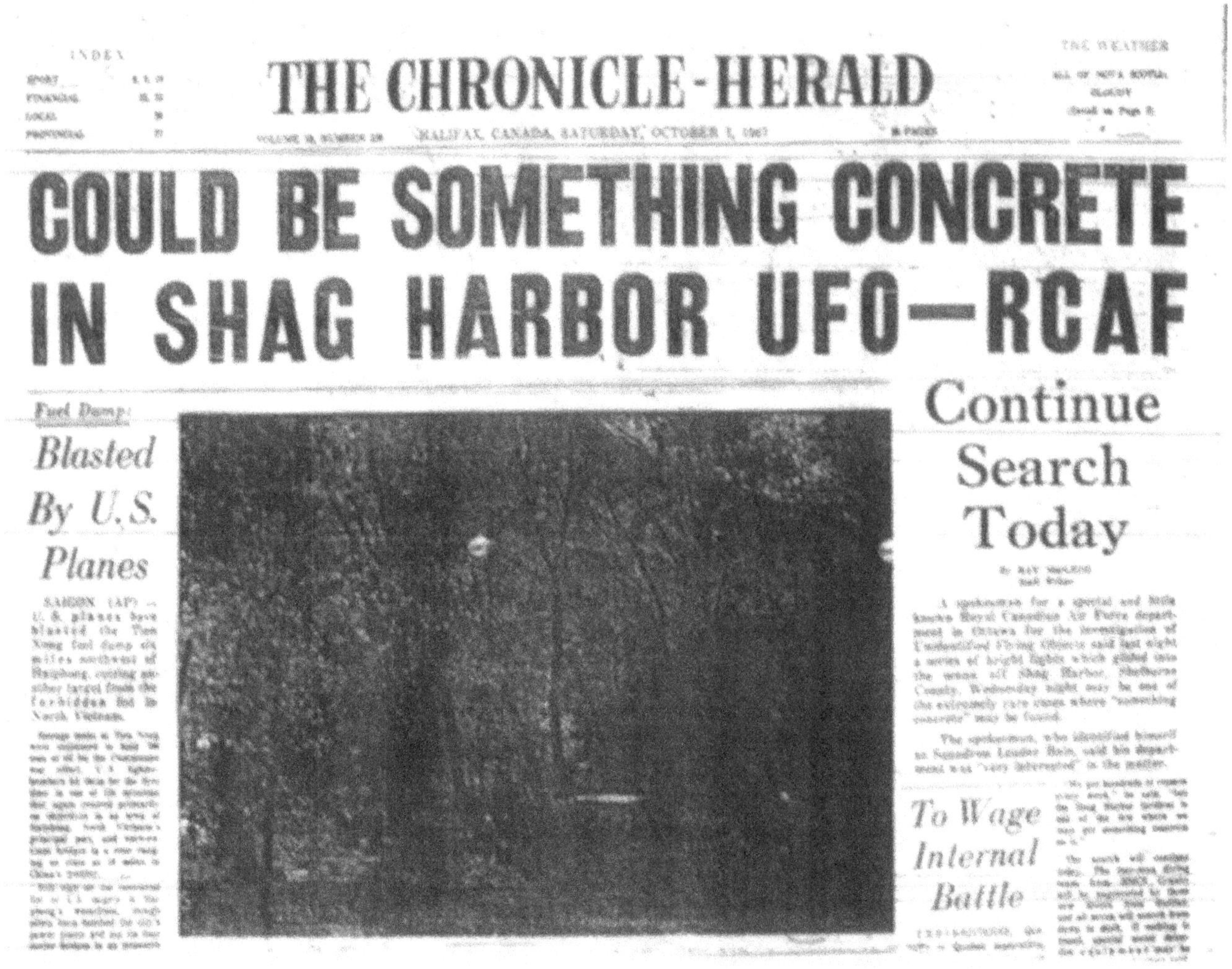
THE CHRONICLE-HERALD

HALIFAX, CANADA, SATURDAY, OCTOBER 7, 1967

COULD BE SOMETHING CONCRETE IN SHAG HARBOR UFO—RCAF

Fuel Dump Blasted By U.S. Planes

SAIGON (AP) — U.S. planes have blasted the Tien Nong fuel dump six miles southwest of Haiphong, cutting another target from the forbidden list in North Vietnam.

Continue Search Today

A spokesman for a special and little known Royal Canadian Air Force department in Ottawa for the investigation of Unidentified Flying Objects said last night a series of bright lights which glided into the water off Shag Harbor, Shelburne County, Wednesday night may be one of the extremely rare cases where "something concrete" may be found.

The spokesman, who identified himself as Squadron Leader Bain, said his department was "very interested" in the matter.

To Wage Internal Battle

Figure 10 Chronicle-Herald News article

spotted on sonar there, and Naval vessels were positioned over it. After a couple of days, the military was planning a salvage operation, when a second UFO joined the first. Common

belief at the time was that the second craft had arrived to render aid to the first.

At this time, the Navy decided to wait and watch. After about a week of monitoring the two UFOs, some of the vessels were called to investigate a Russian submarine which had entered Canadian waters. At this point, the two underwater craft made their move. They made their way to the Gulf of Maine and putting distance between themselves and chasing Navy boats, they broke the surface, and shot away into the skies.

An excellent book on the incident is titled:" Impact to Contact", The Shag Harbor Incident By Chris Styles and Graham Simms.

The Brushy Creek Incident, Piedmont, Missouri

Beginning in February 1973 the Piedmont, Missouri police department received over 500 calls from area residents reporting UFO sightings in the Brushy Creek area. Brushy Creek which is located in southeastern Missouri, borders Piedmont to the north and the town of Mill Spring, nine miles to the south. The area is best known by Missouri residents for its two recreational lakes which are Clearwater just 7 miles west of Piedmont, and Wappapello about 40 miles southeast.

One of the most documented of these incidents happened on February 21, 1973 when Piedmonts Clearwater High school basketball coach Reggie Bone along with five members of his team (the fighting tigers) were returning home from a heartbreaking playoff game defeat losing by seven points to Richland High School in Essex, Missouri which is located 80 miles southeast of Piedmont. The group was returning home traveling on highway 60 near Ellsinore, Mo. about 20 miles south of Piedmont when coach Bone who was driving, noticed a "bright shaft of light beaming down out of the sky." As they continued traveling the dark winding hilly road the group spotted a strange rotation of lights low in the sky. They continued their journey home getting a view of the lights through the wooded hills from time to time wondering what they were seeing. Coach Bone had only driven a few miles further to highway 49 when basketball player Randall Holmes shouted "Look, there's that thing we saw on highway 60." The coach pulled the bus over and the group of six got out on the side of the highway to have a look. The strange lights they had been watching were attached to an object in an open field about 200 yards away. The object was hovering about 50 feet above the ground and had four lights that looked like rotating portholes. The lights were red, green, amber,

and white about three or four feet apart all in a row. One of the team members Cary Barks who was listed in the "Who is Who of American High School Students" because of his academic excellence, stated that the group stood at the edge of the field watching the craft for 5 to 10 minutes when it suddenly without any sound, went directly up in the air and disappeared over a ridge.

About a half hour later at approximately 10:00 p.m., the object or a similar object was seen in Mill Spring by Mrs. Edith Boatwright who reported seeing a strange object flying low over her barn. Mrs. Boatwright was lying on her bed not quite asleep when she saw a flashing light through her window. Since she lives near the highway, she thought there was a problem on the road, so she got out of her bed to have a look. Looking over her curtain she noticed a craft flying low just clearing the utility wires. The craft was in a horizontal position and made a very quiet noise like a whoosh slowly and evenly. Mrs. Boatwright stated that when the craft changed into a vertical position it made a louder noise like a quiet motor "pulling", and that it didn't have any chopper blades on the top, just some rotary blades in front where an umbrella like part extended up. The craft was about 30 or more feet long with a very beautiful light-colored body and a darker tail. It was flying about 200 to 250 yards from her window below the tops of the Oak Trees. Mrs. Boatwright watched the craft for one to two minutes and then lost sight of it.

Edith Boatwright reported her sighting of the strange object outside of her window because she had heard about what coach Bone's team had seen the same night just minutes before. What is very important about this incident is the fact that the Reggie Bone's basketball team was returning home from a playoff game they had lost. Sports are very important in the mid-west, especially small-town sporting events. Losing a playoff game meant a quiet, solemn ride home. Even our local St. Louis Cardinal Baseball team manager had a popular saying with the fans. An out-of-town win for the Cardinals was referred to as a "happy flight" home. Had Reggie Bone's team returned home as winners the ride would have been filled with laughter and excitement and the lights in the trees may never have been seen by anyone on the small bus. Piedmont's most documented UFO sighting of 1973 could have been easily missed. Without the teams sighting report Edith Boatwright might not have reported her sighting. These two UFO sightings on February 21, 1973, set a chain of events in motion. Brushy Creek residents spent the next few months outside looking up. Had they not, much of Piedmont's 1970's UFO Flap might very well have been missed.

Clearwater Lake, Missouri
March 21, 1973

Figure 11 Clearwater Lake, Missouri Photo: Debbie Z.

Just one month after the "Brushy Creek" sighting, a second major sighting event took place.

Clearwater Lake near Piedmont, Missouri is also one of the sights of the "1973 Missouri UFO Flap" sightings. The lake is located on the 37.1589N latitude line at 90.7685W longitude.

Clearwater Lake is a 1,630 acres lake with a maximum depth of 135 feet. On March 21, 1973, Mrs. Jean Coleman and Mrs. Cathy Leach were crossing the Clearwater Dam about 9:00 p.m. when they saw an object rise out of the lake. They were first alerted by a "red flash" on the lake. Stopping their car, they got out to see blinking lights ascending. Each time a red light flashed; the object got brighter. "We could see it climbing," Mrs. Coleman said. "It looked like the lights were red, white and yellow but there was no sound. We tried to make out the shape but each time the lights went out we could see nothing. We watched it for four or five minutes until it circled out of sight." Ken Johnson, owner of the Piedmont Boat Dock, confirmed the women's story. Shortly before they saw the UFO leaving the water, unnamed campers told Johnson they had seen a "bright light moving right under the surface of the lake." Two divers from the East Side Divers Supply Company of Granite City, Illinois were dispatched to the scene. They made three

Figure 12 Jean Coleman (right) with Debbie Ziegelmeyer Photo: Lynne Mann

attempts to explore the lake for evidence of the underwater UFO, but unfortunately an unusually heavy spring rainfall of seven inches had raised the water level 30 feet above normal. The lake was extremely murky, and the divers found nothing in its depths. Clearwater Lake is primarily used in the summer months as a recreational, boating, skiing, and camping site.

State Representative Jerry Howard organized a town meeting in Piedmont the weekend of March 31st thru April 1st of 1973 to address the UFO subject, and multiple area sightings. Several witnesses and UFO investigating experts were invited including the five Clearwater High School basketball team members. Dr. J. Allen Hynek a consultant on UFOs for the Air Force for over 20 years headed the investigation along with Walter Andrus Director of the Midwest UFO Network which would later be known as MUFON. To conserve time, the meeting was held at the home of basketball coach Reggie Bone.

After several weeks of investigation, the object seen by Coach Bone and his team would be classified by the experts as a UFO and categorized as "nocturnal lights" within 200 yards. Strange lights and craft would continue to be seen over the Piedmont area including hovering over and coming in and out of Clearwater Lake for the next several years, making this one of the nation's most active and documented sights in the 1970s for UFO activity.

UFO sightings continue to be reported to MUFON and to me personally even today. I have several friends who have family living in the Piedmont area who claim to spend clear nights sitting outside watching the stars and the UFOs. Some nights bright lights are seen "dancing across the sky" especially in the vicinity of Clearwater Lake.

Pascagoula, Mississippi October 11, 1973

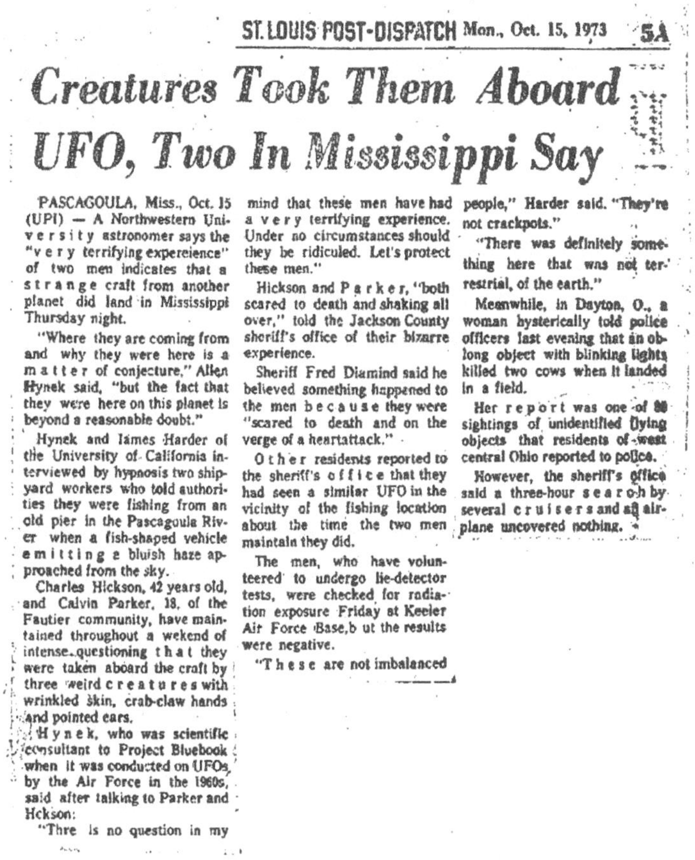

Figure 13 St. Louis Post-Dispatch

Charles Hickson, 45, and Calvin Parker, 18, worked at a shipyard in Pascagoula, Mississippi. The night of Thursday, October 11, 1973, sometime after dark, they decided to go on the west bank of the Pascagoula River where the shipyard is and fish for a while off an abandon dock. (Neither had a watch so they were unsure of the exact time, but additional witnesses set the time at approximately 7:00pm.) They hadn't been there very long when suddenly they heard a buzzing noise. Hickson turned around to see where the noise was coming from and saw a bright blue light approximately 25 to 30 feet away. The light was oval shaped, approximately 8 foot wide and 8 foot high. One end of the blue light opened allowing three creatures to come out, not touching the ground, just floating slowly, a couple of feet off the ground. When the creatures approached the two men, two of the creatures positioned themselves one on each side of Hickson's arms. He stated that he didn't feel any sensation at all when they touched him and was just lifted right off the ground.

Hickson, terrified, remembered being led through the opening the creatures had come out of, into a "room like thing that seemed to glow". He didn't see anything that looked like light fixtures, or any instruments, but stated; "I saw things that I just can't explain. There was a thing in there, it resembled an eye, but it was a big thing like a globe. It just moved all around me." Hickson described the creatures that had taken he and Calvin Parker as having "features of a human being, but it didn't have any hands, they had what looked like pinchers." The creatures were very pale in appearance, approximately 5 foot tall, there were no toes just rounded feet, and either skintight clothes or no clothes at all. There was no sign of hair, and only one of them made some sort of sound. Hickson further described the creatures as not having a human type of nose but "something sitting on a body and a sharp thing coming out about middle-way of the eyes and it looked like an opening to me underneath, and things on the side like ears."

Neither Hickson nor Parker remembered any sound inside the vehicle. Hickson didn't believe he lost consciousness while inside the craft and was unsure how long he and Parker were aboard. There was no sensation of moving and time wise he believed "had to be an hour or so. It had to be that long, but it seemed like an eternity." Neither Hickson nor Parker had any power to move, they were rendered helpless. When asked how they were released, Hickson stated; "they carried me right back out and I was immediately put on

the ground. I felt no pain and I felt normal. Then the vehicle was gone." Hickson stated that he had not seen Parker while on the craft but only after he was released, Parker was sitting on the ground behind him. Hickson stated that Parker was "hysterical and sort of looked like he was paralyzed but he suddenly came to his senses."

Parker recalled the creatures taking him "just like a big magnet drawing me to it. I wasn't on the ground I was off the ground. I don't remember a thing. I just blacked out. I just stood there like I was froze. Then I finally got to where I could move a little bit. It was like a bad dream. I wish it had been a bad dream and it would all be over with. I didn't sleep more than three seconds all night." Parker did not remember being brought back just the "zzzp" sound as the craft left at high speed.

Hickson and Parker were afraid they would not be able to convince people of what they had seen so they waited a while before they went to the Sheriff's Department. Hickson stated that he "wanted a chance to get a hold. I wanted to get the military in on it. I didn't want any publicity and I didn't want any news people, but after I thought about it a while, I figured that was what I should do."

At least a dozen other people were witnesses to the event including a couple who claimed to not only see the blue oval shaped light hovering just above the water but the abduction of Calvin Parker and Charles Hickson from the opposite riverbank. The couple also claimed to have heard a splash in the water and saw human like creatures moving oddly just under the water surface.

The abduction incident was covered by several newspapers including Mississippi Press Register, Washington Star News, Herald-Wig Quincy, Il., Memphis Commercial Appeal,

Springfield Leader and Press, St. Louis Post Dispatch, Dallas Times Herald, The Associated Press, The Blade-Toledo, Ohio, Times Picayune New Orleans, Mobile Press, Mobile Register, The San Francisco Chronicle, The Toronto Sun, and the Toronto Star.

Gulf Breeze, Florida November 1987

The town of Gulf Breeze is unique in that it is an island community, surrounded on three sides by water and on the fourth by a wooded park, the Live Oaks Reservation. A large population of retired military people reside in the area which probably contributes to the apparently broad-minded perspective of the residents. Between 1983 and 1987 many residents of Gulf Breeze and the neighboring communities had been exposed to the UFO subject through Donald Ware's participation in 43 lectures, seven newspaper articles, six television appearances and four radio programs. These appearances touched on three important points; that UFOs have been here for a very long time, some UFOs are controlled by a more advanced intelligence, and a select few in our government know a lot more about UFOs than they will admit. Donald Ware, who lived 44 miles from Gulf Breeze, was notified the day the first UFO picture appeared in the weekly Gulf Breeze Sentinel, so an investigation into these 1987 reports was almost immediate.

On March 13, 1982, MUFON investigated a report similar to those which would follow five years later. A gentleman sighted an oval shaped craft he described as flying low altitude, yawing, and pitching. It moved in a curved pattern, briefly hovered, then with a flash of red disappeared.

November 11, 1987

The most well-known of the Gulf Breeze sightings was reported on November 11, 1987. This began months of reports followed by years of additional sighting reports in the area. This initial sighting of November 11, 1987 came from Edward Walters, a local building contractor. Walters was working late on the evening of November 11, 1987, when he noticed a light coming from his yard. He went to the window to get a better view and saw a glowing object partially obscured by a 30-foot-tall pine tree in his front yard. Walters opened the front door of his house and saw a huge unidentified craft. He immediately ran for and grabbed his Polaroid camera and was able to take several photos. Walters decided to run out into the street for a better view and better pictures. As he did, the craft moved directly over him bathing him in a "blue beam" force field rendering him temporarily paralyzed. Walters described the feeling as follows; "The right side of my forehead felt like it has a knife piercing through to the back of my eye socket." Less than 20 seconds later, Walters describes a "computer like" voice in his head telling him to calm down. Walters shouts "put me down." He then rises from the surface; smells a scent of "ammonia mixed with heavy cinnamon," and a humming sound. The voice instructs him to "S-t-o-p it." Walters shouts "screw you!" Walters hears a voice again in his head, this time female, and in his mind, the sharp vision of a dog, then more pictures of dogs, one after another. Walters owns a dog. He is then released from the force field and falls to the ground, forward onto his knees. The hum subsided and the UFO was nowhere in sight. The MUFON Investigators on this sighting believed that Edward Walters experienced a "failed" abduction.

November 11, 1987

A second witness came forward on November 11, 1987, after seeing a TV report regarding Edward Walters' UFO sighting. The witness had gone to his friend's home to obtain gate keys so that he could remove his boat from a secured area. While driving north on SR 191A (Orliola Beach Road) at about 5:30pm, the witness observed a large object moving slowly from east to west at an estimated distance from this point of 700' - 800'. While continuing on and past Oriola Beach Elementary School, he observed this object pass partially behind distant treetops and into a clearing between other trees. The object was now stationary, and as he turned his truck right (west) he notices 2 Air Force jets, flying directly toward the object in a low-level flight. The witness immediately stopped at the edge of the road to get out of his truck for a closer observation. At this point, the witness was estimated to be 350' - 400' from the object with the object's altitude at approximately 100' - 200'. He described this object as the same one that had been photographed by Mr. Walters which had appeared in the Sentinel and on television. As the jets moved closer, the object moved up about 200' and hovered at a 45-degree angle with more of the underside visible. The witness stated that the object hovered without movement for about 45 seconds. The object then dropped "very fast", straight up until disappearing. At the exact time of the object's departure, a "sun like -flash" was seen. The flash seemed to engulf the whole object. At this point the jets turn right (north) and go over the bay without gaining altitude. The witness said the "flash" left tiny dark spots in his eyes immediately after this observance. He speculates that the flash was caused by the object or by the sun's reflection.

- **November 11th** at approximately 2:30am, the wife of a retired Navy Captain, was awakened by her dog and led outside where she observed a silent, hovering object shine a "pathway" of bluish light onto her dock.

- **November 11th** at 8:15am, several blocks to the north of where the Navy Captain and his wife resided, a witness observed a circular object with two rows of dark spots and a small dome on top hover silently 350 to 400 feet away. As two Air Force jets flew directly toward the object at low level, it quickly moved up about 200 feet, and just before the jets arrived, the object departed straight up "very fast" as a flash of light appeared to engulf the whole object. The jets immediately turned north over East Bay. The witness described the object as being approximately 30 feet in diameter, about 15 feet high with a dull-silver top, light-tan middle portion, and a dark-beige bottom.

November 24, 1987

On November 24, 1987, a woman and her son were traveling east on Highway 98. As they passed the Naval Live Oaks Visitors Center, the woman noticed a lighted object in the sky to her left. It was traveling east, the same speed as her car. She thought it was a helicopter but joked with her son about a UFO. When they reached their destination, the object had stopped and was hovering about twenty feet above the trees across the highway. They stopped in the Santa Rosa State Bank parking lot, which is adjacent to the Villa Venyce entrance, to watch. The craft was disc-shaped, about the "diameter of an octopus ride at the fair", but massive and solid-looking. There was a single row of windows around the middle, with red and white lights alternating in the windows. The craft seemed to be silently spinning. Suddenly, a red ball of light dropped from the bottom of the craft. The craft then took a 90-degree turn to the right, then another 90-degree turn upward and stopped momentarily. It then shot out of sight very quickly at an angle and disappeared. The witness and her son extremely "rattled", decided to go home immediately. The sighting lasted about five minutes. On February 12, 1988, the witness's son and some friends saw the same craft traveling the same route, except in a westerly direction. It would periodically lurch forward very fast and take on the appearance of a streak of white light. As they entered Gulf Breeze from the east, it quickly disappeared.

Multiple 1987 Sighting Reports

In 1987, Florida MUFON received several UFO reports from Florida locations outside the Gulf Breeze vicinity, but these were not documented with photos.

- **January 1st** a woman reported red lights hovering low over highway 6 in north-central Florida.
- **January 8th** a couple reported a large vertical cylinder above their house in Lakeland.
- **June 15,16, and 17th** many witnesses reported UFOs near Ocala. This was followed by a three-hour series of sightings by many witnesses, including two police officers. Strangely, 14 dogs in the neighborhood were reported missing.
- **July 23rd** several people saw six spherical objects make 90° turns over West Pensacola during daylight hours.
- **August 5th** a huge, silent, circular object with colored lights stopped 100 feet above a witness near Gainesville.
- **October 19th** a woman reported a UFO hovering beside a bridge near Arcadia in daylight, close enough to see beings through a window.
- **November 25th** a woman in Destin reported the apparent temporary abduction of her five-year-old son.

The above sighting events of November 11, 1987, and those which followed, are just an

Figure 14

example of the kind of activity that continued around Gulf Breeze for many months. Five original Polaroid photos taken by Edward Walters, were given to the editor of the weekly Gulf Breeze Sentinel, Duane Cook. Walters was known by the editor for his large charitable contributions in support of high school activities and the youth of the community, so Cook decided to publish the pictures. Cook was surprised to learn that his mother had seen the same object a few minutes before the pictures were taken. Personal friends of the MUFON State Director lived in Gulf Breeze. They notified him on November 19th when the pictures were published, beginning a MUFON investigation which continues even today.

Hudson Valley, New York

The Hudson Valley area in New York has been a key focal point for sighting reports for years. In 1987 Dr. J. Allen Hynek and Philip J. Imbrogno wrote a book along with Bob Pratt titled "Night Siege" "The Hudson Valley UFO Sightings". The book focuses on "five thousand responsible people who claim to have seen something." Some claim to have seen an unidentified craft larger than a football field, dark metallic gray, brilliant flashing color lights that formed a "V". The book describes sightings seen by housewives to aviation engineers, from children to professional businesspeople describing something above homes, highways, ponds, and backyards. These sightings were reported from one of the most densely populated suburban areas in the U.S.A. Because of the number of UFO sightings that have been reported in the Pine Bush, New York area, Pine Bush is known as the UFO Capital of the East

Debbie Ziegelmeyer

Coast. Residents and visitors have reported strange sightings in the sky dating back to the early 1960s. The high volume of UFO activity in the Hudson Valley area, has attracted attention from mainstream news and media. MUFON New York continues to receive witness sighting reports in the area and has several on-going investigations.

January 2019

In May of 2019, the MUFON Journal published an article titled "UFO Dumps Object Into New York Lake", by Joseph Flammer, New York MUFON Field Investigator. The article was written as a "call for witnesses to step forward to help solve a potentially dangerous UFO mystery" which is underway in upstate New York. "Project Aries" was formed to "figure out what a black triangular UFO dumped into a lake in the rural town of Tuxedo in January 2019." A report was filed with MUFON in late January describing the incident, claiming there was a second witness to the event.

December 31, 1981

Hudson Valley sightings of unknown craft are thought to have begun on New Year's Eve, December 31, 1981. An off-duty police officer was outside in his yard at his home in Kent, New York, with his family just a few minutes before midnight when in the distance, they noticed a group of red, green, and white lights. The lights were approaching from the south, and at first were thought to be an aircraft. As the object approached, the family noticed that the multicolored lights were not from a plane but coming from a boomerang-shaped craft. The strange craft slowly passed directly overhead at approximately 500 feet altitude, emitting a faint humming sound with alternating lights from multicolored to all-white displaying

different patterns. Several additional witnesses reported seeing this strange craft including witnesses on nearby Interstate 84 who stopped their vehicles along the roadside to watch. One of the motorists was 55-year-old Edwin Hansen, who described the massive boomerang or V-shaped craft as being "big enough to block out the stars and could be seen to project some sort of light beam to the ground". Hansen would also claim to have made what he believed to be some type of telepathic contact with the craft, filling his mind with strange images and thoughts that were not his own. He also had the sensation that something was "reading his mind and telling him not to be afraid".

February 26, 1982

On February 26, 1982, Monique O'Driscoll was driving with her daughter when they saw a strange light in the sky that seemed to be hovering over a frozen lake. O'Driscoll stopped her car and described seeing a "large V-shaped object with numerous red, blue, and amber lights dotting it". She stated that the underside of the craft was crisscrossed by some sort of metal beams. The object seemed to approach them, giving them quite a scare, before departing into the night sky. This same object was seen by several other witnesses who overwhelmed the local police department with panicky calls.

March 17, 1983

An unidentified craft with the same description as reported before, was seen by witnesses in Brewster, New York on March 17, 1983. The witnesses included the Deputy Clerk for Putnam County Dennis Sant, who saw it from his home. Sant described what he saw as being "a very large object, the structure of it was very dark gray, metallic, almost girder-type looking. The object seemed to be very silent. The lights were iridescent, bright, they stood out in the sky and three-dimension. It looked like a city of lights. It just hung in the sky, all brilliant colors". He and his family followed the moving object around to his backyard view but at that point, he stated that "a feeling of fright came upon me. Thoughts started to flood my mind, thoughts of the craft touching ground, thoughts of an encounter with an alien being. Thoughts of being abducted. All types of fearful thoughts started to enter into my mind". Once again, the unknown craft was described by several additional witnesses as being a large boomerang-shaped craft, approximately 131 feet long, greater than 330

feet wide, with a multitude of bright red, green, amber, and white lights along its fuselage, traveling silently overhead.

March 24, 1983

On March 24, 1983, in the areas of Yorktown and New Castle the local Police began getting numerous calls from people seeing unexplained objects in the sky. The witnesses were describing the same V-shaped lighted craft mostly described as being "as large as a football field". This time however, several witnesses reported seeing smaller objects darting through the sky, in some cases, in two in different locations at the same time. It seemed that this time, there was more than one unknown craft being reported. Several police officers were among the witnesses including Officer Andi Sadoff of the New Castle Police. Officer Sadoff explained what he saw in the following report: "I was working a 4:00 p.m. to midnight tour and assigned to set up some radar to look for speeding cars and I looked up into the sky and saw a series of lights. At first, I thought it was a plane, it was quite a distance, quite far away, but it was, it was really quite large. As I recall, there were mostly white lights, but there were green lights also. It was alternating green and white lights. It approached my vehicle and it stopped and it seemed to hover. And I'm looking at this thing, thinking what is it? I wasn't afraid. I was just amazed. I was in awe of it. I didn't know what it was. The only thing that I recall the most is I was amazed that there was no noise. There was no humming. There was no engine, there was no sound. It was absolutely silent".

A similar report was filed by another witness at exactly the same time. The witness was a senior manager for IBM by the name of Ed Burns. Burns was approximately 10 miles away from Officer Sadoff, when his radio started acting oddly. Burns stated, "Out of nowhere, I got a lot of static on the radio. I thought maybe I was on the wrong number, and then I went over to turn the dial again and that's when I looked up and saw this craft. It was a triangular ship. And the back had to be as large as a football field at least. And there was no noise. I'm not into astronomy... but what I had witnessed that night was not from this planet".

The large unidentified craft generated sighting reports up and down the Hudson Valley startling motorists and numerous police officers. The following evening the strange craft was seen by numerous people who again described the craft as being large, "as big as a football field" or "as large as an aircraft carrier", V-shaped or triangular with lights that studded the exterior. Some witnesses claimed that the craft would "shoot beams of light", "spin around like a top", "shoot away at fantastic speeds", "display patterns of colors with their lights", "detach what appeared to be smaller craft from the main body", or "wink in and out of existence". Jim Booke, a biomedical engineer, claimed he saw the craft silently circle over the Croton Falls Reservoir dropping what seemed to be lighted probes into the water. The craft then began shooting a red beam of light into the reservoir. Several witnesses reported seeing the strange craft hover over bodies of water.

The magnitude and frequency of these sightings generated interest from the press showing up in major newspapers nationwide. UFO researchers Philip J. Imbrogno and J. Allen Hynek arrived in the area to begin interviewing many of the witnesses including those from December 31, 1981, the first reported sighting in the area. A phone hotline for witnesses to the events was set up which almost immediately brought hundreds of calls by those who had sighting experiences. The floods of hotline calls and those to the police, brought the estimate of witnesses to the March 24th event to be in the thousands. The

majority of the witnesses were from the Taconic Parkway area, many who were found to be reliable upstanding citizens. All of the witnesses described the same enormous boomerang shaped craft. Imbrogno and Hynek were convinced that something that was definitely out of the ordinary was occurring in the area, but the local police were not. According to law enforcement authorities, the events were pranks being pulled by normal ultralight planes or helicopters flying in formation in such a way as to seem more mysterious. The Federal Aviation Administration agreed with this theory, as did officials from nearby Stewart Air Force Base. A report from air traffic control specialist Anthony Capaldi, who had also seen the phenomenon, concluded that what people were seeing, and reporting were planes flying in formation. He would state; "The first time I observed the formation, it looked a little peculiar. And from our vantage point in the tower, they just appeared to be just one big light because they were flying in tight formation. I don't think if this formation flew over an individual's head at a thousand feet that there's any way you could mistake it for anything but the formation flying, due to the sound of the aircraft engines. And I imagine that at a thousand feet, you could really determine that it's aircraft".

April of 1983

In April of 1983 a HOAX event was attempted but failed when most people who saw this formation immediately recognized them as planes in formation. Many of the witnesses insisted that they knew the difference between planes and what they had seen and refused to believe that the objects were normal aircraft flying in formation at high altitude too high for the sound of their engines to be heard. Except for the April hoax, no one had ever been caught in the act of attempting to perpetrate a HOAX in the area, there were no suspects, and no reason to spend that much time, effort and the expense of fuel needed to fly the "HOAX event" planes. Ultralight planes would unlikely be able to remain silent and maneuver in flight in such aerial activities as reported by witnesses or hold that type of formation. Imbrogno and Hynek concluded that planes in formation were not an acceptable explanation.

March 31, July 12, 19, and 24, of 1983, and on March 25, 1984

Sighting reports by witnesses of the same large unknown triangle shaped craft, continued on March 31, July 12, 19, and 24, of 1983, and continued to March 25, 1984. Each one of these sighting events generated hundreds of witness reports including a report from a crew at the Indian Point Nuclear Plant who claimed that the massive craft had hovered over one of the nuclear reactors and caused the facility's security systems to shut down. The shut-down tempted the supervisor to have his men open fire on the strange craft, before it quickly shot off into the night sky. The sighting reports seemed to stop in 1986. Imbrogno and Hynek estimated that more than 5,000 people had seen these craft and would later write a book titled "Night Siege: The Hudson Valley UFO Sightings". Imbrogno and Hynek were also able to find certain idiosyncrasies among the data they had collected, such as the sightings being clustered very tightly and seemed to fall on Monday, Thursday, and Sunday nights to a disproportionate degree. Although mass sightings such as those reported between December 31, 1981, and 1986 seemed to have stopped, sighting reports of unknown craft hovering over bodies of water in the Hudson Valley area have continued.

2018

January through December of 2018 the Hudson Valley had numerous people witness unusual things happening in the skies above.

The National UFO Reporting Center had at least 11 UFO sighting reports in the region.

- **January 12th, 2018, just before 4p.m.** The first sighting in the Hudson Valley in 2018 occurred and was described as a "sphere of light" which was spotted over the Hudson River in Poughkeepsie. The reported sighting lasted one minute.
- **February 2nd at 4:45 p.m.** The next sighting occurred and was described by witnesses as a fast-moving oval object seen over Pawling.
- **March 24th** Unknown flickering lights were reported in the skies over Bloomingburg in the early hours of the morning.
- **June 8th** Several orange oval objects were seen floating across the skies over Hudson for more than 90-minutes.
- **June 16th** Orange objects were seen in the night sky over White Plains.
- **September 15th and 16th** Middletown, New York was the only town to make the list of sighting reports twice, with reports being filed of glowing fiery balls of light seen in the skies above.
- **November 12th** An unknown craft with strange flashing lights was seen flying over Woodstock.
- **November 30th** A large dark unknown craft was seen floating above I-84 below New Paltz.
- **December 7th** A blue star-like object was seen behind an airplane over Westtown following it from behind before disappearing.
- **December 26th** A large disc with yellow and red blinking lights rotating around its center, was seen over I-84 in Fishkill.

New York MUFON State Director Sam Falvo, and New York MUFON Assistant State Director Chris DePerno, have been investigating UFO sighting reports in the Hudson Valley for years, stating that the area is "a hotbed of activity". There is an on-going 2019 investigation into a waterway located in Orange County, where witnesses claim to have seen a triangle of lights hovering over a body of water. DePerno states that credible sources reported seeing barrels dumped from an unknown object into the water before disappearing into the night sky. I was asked to join a team of scuba divers in search of these barrels and other unidentified materials including possible related evidence, but permission from the local authorities for water entry by scuba divers has delayed this investigation. Water samples were taken in September 2019 and analyzed by Lynne Mann, Chief Lab Technician for the MUFON Analytical Laboratory, but proved to be inconclusive due to the length of time between the incident and the date of testing. This investigation and new sighting reports and investigations of unknown craft seen in the area continues to date.

Fall-Winter 1989 Tillamook, Oregon Area

On September 27, 1989, at 9:30pm, a greenish-blue object was reported streaking across the sky passing low over the horizon by multiple witnesses which at first was mildly dismissed as a passing meteor. The following evening of the 28th, the Oceanside area had multiple witness reports to local police departments of multicolor balls of light in formation observed flying and hovering overhead.

Beginning at approximately 8:30pm on the 28th, one witness reported that she and many of her neighbors observed circular lights moving around the sky. As the lights moved closer, they noticed the lights were actually a craft which they described as being circular, the size of a commercial airliner, with three lights in front, and one on either side. The craft moved north towards Leyden, hovered for 2 ½ hours, then at approximately 11:00pm dropped down behind the mountain out of sight.

Local Radio station KTIL received between 35 and 40 calls on September 28th, of a strange cluster of lights seen in the sky over the Pacific Ocean being viewed from Oceanside. Witnesses reported seeing a cluster of 6-8 lights "hemispherical shaped" pointing down over the ocean similar to "waterdrops on the underside of a plane". Some of the lights overlapped others with the brightness of the lights or "hue" being orange and more intense on the outer edges. The cluster would emit a beam of light horizontally terminating in a single bright "spherical-shaped" light. The beam distance equaled three times the base of the length of the cluster. The beam would then collapse creating some sparkle, then suddenly all of the lights would turn off. The cluster would then appear again in a different location in the sky and the sequence would begin again. This sequenced event description occurred several times over the next several minutes and was witnessed by Oceanside residents looking out over the Ocean towards a clear horizon. Other witnesses described a bright light with flames shooting out of it, disappearing, and then reappearing six times, each time getting brighter as objects shot out from it.

Between September and November of 1989, multiple sighting reports of unknown aircraft and "strange lights" came into local police offices, radio stations, newspapers, and UFO reporting groups including MUFON (Mutual UFO Network). MUFON received reports from Aloha, Tillamook, Portland, Rockaway,

Oceanside, and several other locations with estimated sighting reports nearing fifty. Local residents complained to MUFON that their reports to the local Coast Guard and radio stations were falling on death ears. They felt that local authorities did not seem interested in investigating these sightings witnessed by so many and would not direct them to an organization with people who would. Local authorities may have been as baffled by these events as the witnesses or may have been told by some unknown authority to "play them down".

October 9, 1989

In October, the UFO sighting reports from multiple witnesses and multiple areas, turned to a new unsolved mystery with the reports of two missing fishermen. On October 9th fisherman Oscar Schultz left port from Charleston, Oregon on the "Gloria L.". The Gloria L., which was a 28-foot salmon troller, was captained by Oscar Schultz, a seventy-year-old commercial fisherman with over thirty years of experience. The fisherman left port at 6:00am, a sunny day with calm seas, temperatures in the 70's, seemingly a perfect day at sea. When Schultz failed to return to port, a search for the missing fisherman was launched. Three days later on October 12th, Schultz's boat was found by the Coast Guard drifting forty miles off the mouth of the Umpqua River with no one on board. The "Gloria L." was spotted by the crew of a falcon jet from the Coast Guard Station in Astoria. A Coast Guard helicopter was dispatched to the scene where a rescue diver was lowered on to the boat to investigate. The diver who found the boat unmanned with no sign of foul play, ran the abandoned boat back to Coos Bay. A statement was released by the Coast Guard, that Schultz "must have had a heart attack and had fallen overboard".

Fisherman Jerry Buck also left port on October 9th from Newport, Oregon on his boat the "Acie-O. On October 13th Buck's boat was also found drifting approximately 10 miles from where Schultz's boat was discovered just the day before. Buck set out to sea during calm, clear sea conditions aboard his fishing boat the 36 foot "Acie-O". When Buck, like Schultz did not return to port, a search was launched for him as well. The Coast Guard cutter "The Citrus", discovered the "Acie-O" fifty miles west of Reedsport in three-foot seas. A crew was put aboard who found the boat abandoned, in gear, engines running, the boat steaming in circles, with a pot of beans cooking on the stove. A 2,500 square mile search was initiated but called off at 6:35pm due to darkness. Tom Curry a Newport Marine Surveyor stated that Buck fished for salmon and crab, usually fished alone, and doubted if a wave had swept Buck overboard because the weather was clear, and the sea was calm.

The Coast Guard helicopter had found Schultz's boat the day before but somehow missed the Acie-0, maybe because they were 10 miles apart. Weather was proven not to be a factor, so what happened to the missing fishermen? Schultz's boat was at sea for three days and Buck's for four, yet there were no fish in the holds of either boat. Articles appeared in local newspapers with reports from residents living off the coast, of observed strange lights seen over the Ocean during the week of the 9th to the 13th.
Neither fisherman was ever found.

Roache Harbor, Washington October 1990 11:30am

Little Pearl Island is a small 4-5-acre Island and part of the San Juan Island parks in the Roache Harbor area. The witness and a friend reported seeing a large sphere-shaped craft hover over Little Pearl Island. The pair were standing at the edge of a 25 ft. bluff over-looking the

Island. The craft was approximately 50 feet in diameter and as tall as a "2 story building". It circled the island for about 5 minutes, hovered about 25 feet over the water for about 25 minutes then descended into the water. The craft remained underwater for 10-15 minutes before rising upward into the NE sky at an extremely fast rate of speed and out of sight.

What Is in The Baltic Sea?

The Baltic Sea anomaly is a 60-metre circular rock-like formation on the floor of the Baltic Sea, discovered by Peter Lindberg, Dennis Åsberg and their Swedish "Ocean X" diving team in June 2011. In March 2014, the Ocean X team got a closer sonar image look discovering right angles, walls with absolutely smooth surfaces and cavities-like corridors inside the object as well as something resembling a staircase.

Rumors about a Baltic Sea UFO began circulating after underwater explorers Denis Asberg and Peter Lindberg of Ocean X returned from an expedition in the summer of 2011 with an unidentified sonar image. The Ocean X Team's main focus has been to search for hidden treasures and historic artifacts. The Ocean X Team started their wreck career by finding an American B-17 bomber in 1992, off the Swedish east coast.

Lying approximately 300ft below the surface measuring 26 foot tall, Ocean X explorers Denis Asberg and Peter Lindberg found the mysterious craft using a side-scan sonar. As told to TV4 in Sweden by Denis Asberg and Peter Lindberg; "There is a mountain, it could be 20-25 meters tall with a canyon in the middle, and below it there is a lot of loose rocks. There are two trails, one that leads to this large circle, then there is another trail that leads to the second object which is about 200 meters from the round circle. We found it at the same time as the circle, but we had such focus on the circle that we have not had

enough time to look at the other object. It is not circular. We were really surprised and puzzled thinking, what is this we have found here. This is not a wreck."

The two explores wanted to do some research on the area and decided to keep this discovery totally quiet. They contacted geologists and marine biologists who said they had never seen anything like this. Part of the structure looked like a staircase and appears to lead to a dark hole with a second structure found nearby. One thing is certain, despite looking like a huge rock, scientists are saying the structure appears to be made of metal.

The mysterious side effect of this object is the EMF (electromagnetic field) it generates. Stefan Hogerborn, a professional diver with Ocean X, stated that "anything electric out there, and the satellite phone as well, stopped working when we were above the object, then when we got away about 200 meters, it turned on again." Sonar discovered drag marks behind the object and geologist Steve Weiner claims that according to his tests, the structure is not a geological formation. The Swedish diver documented a 985-foot flattened out "runway" leading up to the object, which seems to imply that it skidded along the path before stopping.

Hogerborn stated that the object is made of "metals which nature could not reproduce itself". Lindberg exclaimed, "I have been the biggest sceptic, I was kind of prepared for finding just stone. For me it has been an amazing experience. Whatever it is, it is something we do not usually find in nature sitting in the dark cold depths of the Baltic Sea." Dennis Asberg stated; "I am one hundred percent convinced and confident that we have found something that is very, very, very unique."

Ocean X returned to the site the following year to further explore and document the object, but due to mysterious EMF effects, Ocean X was not able to get a good sonar image of the submerged anomaly. The leader of the team Peter Lindberg claimed he had found a second object; "I confirm that we have found two anomalies. We did find the other anomaly approximately 200 meters (about 219 yards) from the circular find at the same sonar run." Lindberg, however, did not release the sonar image of the second object explaining; "We decided not to expose that anomaly so much because there is a lot of disturbance on the sonar image when we passed it, so it's very blurry. We can see it's something but to an untrained eye it might just look like 'pea soup.'"

An explanation of the sonar image problems was given by Hanumant Singh, a researcher with the Woods Hole Oceanographic Institute, as being use of inexpensive equipment and the sonar equipment used was not properly calibrated. Neither of the sonar images provided a reliable look at the Baltic Sea anomaly. With only a single blurry image and little information, many people speculated the object at the bottom of the Baltic Sea could be a UFO. Artist Hauke Vagt created an illustration of the Ocean X team find described by diver Peter Lindberg as being "the closest so far" to what they believe lies mysteriously below the surface in the deep dark waters of the Baltic Sea.

An update on the Baltic Sea anomaly comes in an article by Sebastian Kettley published on February 22, 2019. The article is an interview with Ocean X diver Stefan Hogeborn who states that the waters around the anomaly were the coldest he has ever dived in. He said, "I've never experience temperatures like this while diving." The rocks form a ring, just like somebody put them there. Stefan Hogeborn,

Ocean X, described a rocky dome with a straight edge and protrusions with a shape some have" likened to the Millennium Falcon spaceship from Star Wars." Hogeborn stated that the anomaly closely resembled concrete rather than natural rock and appeared as if somebody "molded it before it turned into stone". He also discovered a formation of rocks arranged in a circle on the anomaly, which he stated, "eerily looked as if they were left there by someone". Hogeborn stated that this was one more element of the puzzle, which "greatly baffled him".

Detailed Description of Underwater Object

Completely circular plate-like exterior:
- 180 meters in circumference.
- Object is 200ft across mushroom shaped comprising a thick pillar rising 8 meters out of the seabed with a 4-meter-thick dome on top (like a mushroom shape) rising a total of 12 meters (approx. 40 feet) high above the surrounding seabed.
- Cavities-like corridors inside the object.
- Straight and smooth walls in certain areas with many right angles.
- There are visible formations on top of the object, which are set at a 90-degree angle and look like passageways or walls, as well as something that looks like it could be a staircase.
- Stone circles, like "fireplaces", of hard black "almost petrified" burnt looking stone each a few inches in diameter, like 4 or 5 pearls in a necklace in various arrangements on top of the dome.
- Spherical object nicknamed "The Meringue" is 4 meters wide and sits on top of the object.
- Twenty-five-centimeter hole on top of the object, it's not known yet where it leads or what if anything is inside.

- Long runway or "skid marks" leading to object point north.

"When we went out and saw the walls which were straight and smooth, it was frightening, as in a science-fiction film" – Dennis Åsberg – Ocean Explorer Co-Founder

The Ocean X team are continuing to investigate the object, while exploring the floor of the Bothnian Sea between Sweden and Finland. Ocean X is expected to return to the wreck to continue investigating their mysterious discovery.

Chapter 4

Mysterious Triangles

The mysteries of particular triangular shaped land masses and ocean or sea areas, have been discussed for centuries. There is a common belief that these "Triangle" areas are located in vile vortices. This means that these areas have extreme electric, magnetic, and electromagnetic anomalies in addition to energy vortexes, which are electromagnetic currents. Other places said to have these unique circumstances are the north and south poles, Easter Island, Stonehenge, and the Pyramids of Egypt. It has been speculated that these monuments were built in these locations purposely with knowledge of a vortex. Energy vortexes are said to cause physical effects such as altering ones mental, physical, and emotional health, as well as causing hallucinations, visions, confusion, and disorientation. In addition, people living in these areas have shown signs of possessing "healing powers". Electrical instruments in the areas of these "Triangles" can malfunction, such as compasses, vehicles, airplanes, ships navigation systems, watches, phones, and many other types of mechanical instrumentation. Vortexes are also believed to open doorways to the spiritual world or another realm which could be a probable explanation as to why so many people disappear in these regions.

James Allen Hynek in his book "Close Encounters of the Second Kind" Chapter 9, discusses interference in electrical circuits, causing car engines to cease temporarily.

Some of these EMF (electromagnetic field) effects listed were:
- Radios to cut out
- Car lights failure
- Car batteries to overheat

Aircraft malfunctions also listed as possibly being caused by EMF effects were:
- Electrical
- Radio communication
- Unexplained Loss of control
- Wings torn from fuselage

Due to the numerous incidents of unexplained downed aircraft over the Great Lakes region, The Federal Aviation Administration published the following manual.

The Aeronautical Information Manual (AIM)

Aware of the curious incidents over the Great Lakes, the Federal Aviation Administration instituted a special "Lake Reporting Service." Pilots on Great Lakes overflights make continuous reports to ground stations. A 10-minute delay in such a report automatically launches a search-and-rescue operation. This service has saved many lives that would have been lost to ordinary accidents, but the high incidence of unusual disasters has remained unaffected.

e. Lake Reporting Service.
Cleveland and Lansing FSS Radio Sectors provide Lake Reporting Service on request

for aircraft traversing the western half of Lake Erie; Green Bay, Kankakee, Lansing, and Terre Haute. FSS Radio Sectors provide Lake Reporting Service on request for aircraft traversing Lake Michigan.

Bermuda Triangle, Atlantic Ocean

The Bermuda Triangle, also known as the Devil's Triangle, is an area in the western part of the North Atlantic Ocean and part of what is referred to as the Caribbean. The exact boundaries are not clearly defined, but common assumed triangle points are Bermuda, Miami, and Puerto Rico. A number of aircraft and ships are have disappeared under mysterious circumstances over the years including a group of 5 TBM Avenger torpedo bombers. The area of the Bermuda Triangle is amongst the most heavily traveled shipping lanes in the world. Ships traffic frequently crosses through the area for ports in the Americas, Europe and the Caribbean islands. Cruise ships and private pleasure boats regularly sail through the region, and both commercial and private aircraft routinely fly over this area of the Atlantic Ocean. Documented evidence indicates that a significant percentage of the incidents could be related to "rogue waves" but evidence of electrometric fields in the skies above and mysterious underwater lights prove that this area is far from business as normal. Ships have mysteriously disappeared with no explanation, pilots have reported incidents of "wormhole" travel describing incidents as either loosing or gaining several unexplained minutes of time, and sailors have disappeared from their boats with no explanation, never to be found.

Bermuda Complex

The Bermuda Complex is located in the open ocean east of Bermuda clockwise to the south of the island. The Bermuda Complex consists of the following non-instrumented warning areas and Operating Area (OPAREA):

Warning Area 3014 (W-3014)
Warning Area 3015 (W-3015)
Warning Area 3018 (W-3018)
Bermuda OPAREA

Greater than fifty ships and over twenty aircraft have disappeared into oblivion in this area known as the Devil's Triangle, Bermuda Triangle, and Hoodoo Sea, most of which have never been found. Very few distress calls from these vessels and aircraft have been sent or received, and very little, debris if any is ever found. Even with these numerous mysterious disappearances, The U. S. Board of Geographic Names does not recognize the area known as The Bermuda Triangle as an "official" name, and does not keep official files on lost vessels, aircraft or human lives taken. GlobalSecurity.org

Flight 19 By Lt. Comdr. Horace Bristol, U.S. Navy photo 80-G-427475 - This media is available in the holdings of the National Archives and Records Administration, cataloged under the National Archives Identifier (NAID) 520770., Public Domain, https://c

The Disappearance of Flight 19, December 5, 1945

One of the most famous Bermuda Triangle mysteries is the fate of Flight 19 which was on a routine navigation and combat training exercise called "Navigation problem No. 1", a combination of bombing and navigation. Other flights had completed this task or were scheduled to that day. Flight leader Navy Lieutenant Charles Carroll Taylor, had approximately 2,500 flying hours, mostly in aircraft of this type. The student pilots Marine Captains Edward Joseph Powers and George William Stivers, Marine Second Lieutenant Forrest James Gerber and USN Ensign Joseph Tipton Bossi, had recently completed other training missions in the area where this flight was to take place. Each aircraft was fully fueled, but during pre-flight checks, it was discovered that the aircraft were all missing clocks. Navigation of the route was to teach "dead reckoning principles", which involved calculating, and elapsed times. The apparent lack of working timekeeping equipment was not a factor for concern as it was assumed each man had his own watch. Takeoff time was 14:10, weather at NAS Fort Lauderdale was described as "favorable, and sea state was moderate to rough", with Lt. Taylor supervising the mission.

The exercise involved three different legs, which was normally four, with a heading 091°due east until reaching Hen and Chickens Shoals where low-level bombing practice was carried out. The flight was to continue on that heading for another 67 nautical miles before turning to a course of 3 46° for 73 nautical miles taking the group over the Grand Bahama island. The next scheduled turn was to a heading of 241° to fly 120 nautical miles which would end, turning them left to return to NAS Ft. Lauderdale.

The practice bombing operation was completed by Flight 19 because radio conversations between the pilots were overheard by base and other aircraft in the area. At approximately 15:00, a pilot requested and was given permission to drop his last bomb. Forty minutes later, another flight instructor, Lieutenant Robert F. Cox who was lining up with his group of students for the same mission, received an unidentified transmission from an unidentified crew member asking for a compass reading. The unidentified pilot transmitted, "I don't know where we are. We must have got lost after that last turn." Lieutenant Cox transmitted; "This is FT-74, plane or boat calling 'Powers' please identify yourself so someone can help you." The response was a request from the others in the flight for suggestions. FT-74 tried again, and a man identified as FT-28 Lieutenant Taylor came on. "FT-28, this is FT-74, what is your trouble?" "Both of my compasses are out", Taylor replied, "and I am trying to find Fort Lauderdale, Florida. I am over land but it's broken. I am sure I'm in the Keys, but I don't know how far down, and I don't know how to get to Fort Lauderdale." FT-74 informed the NAS that aircraft were lost, then advised Lieutenant Taylor to put the sun on his port wing and fly north up the coast to Fort Lauderdale. A transmission from base operations asked if the flight leader's aircraft was equipped with a standard YG (IFF) transmitter, which could be used to triangulate the flight's position, but there was no reply by FT-28. At 16:45, FT-28 radioed: "We are heading 030 degrees for 45 minutes, then we will fly north to make sure we are not over the Gulf of Mexico". Lt. Taylor indicated that his transmitter was activated but IFF could not be picked up. Lt. Taylor was advised to broadcast on 4805 kHz but this order was not acknowledged. He was then asked to switch to 3000 kHz, which is the search and rescue frequency, but Lt. Taylor replied "I cannot switch frequencies. I must keep my planes intact." At 16:56, Lt. Taylor was asked to turn on his transmitter for YG if he had one, but he did not acknowledge. His transmission overhead was" Change course to 090 degrees (due east) for 10 minutes." About the same time someone in the flight said "Dammit, if we could just fly west, we would get home; head west, dammit." They may have been saved had they turned west, but instead, followed Lt. Taylor's orders as expected.

The weather began to deteriorate at this point and radio contact became intermittent. The five aircraft by this point, were most likely more than 200 nautical miles out to sea east of the Florida peninsula. Lt. Taylor radioed "We'll fly 270 degrees west until landfall or running out of gas". He requested a weather check at 17:24, but by 17:50, several land-based radio stations had triangulated Flight 19's position as being within a 100 nautical mile radius of 29°N 79°W. This put Flight 19 as north of the Bahamas several nautical miles off the coast of central Florida. At 18:04, Lt. Taylor radioed to his flight "Holding 270, we didn't fly far enough east, we may as well just turn around and fly east again". The weather by then, had deteriorated even further and the sun had set. At approximately 18:20, Lt. Taylor's last message was received, "All planes close up tight ... we'll have to ditch unless landfall ... when the first plane drops below 10 gallons, we all go down together." By now it became obvious that Flight 19 was lost, and all air bases, aircraft, and merchant ships were alerted.

A few months later, a 500-page Navy board of investigation report was published.
The following observations were made:
- Flight 19 leader Lt. Charles C. Taylor had mistakenly believed that the small islands he passed over were the Florida Keys and that his flight was over the Gulf of Mexico.
- He believed that heading northeast would take Flight 19 to Florida.
- It was determined that Lt. Taylor had passed over the Bahamas as scheduled, and he did in fact lead his flight to the northeast over the Atlantic.

- The report also noted that some subordinate officers did likely know the approximate position of Flight 19 as indicated by radio transmissions stating that flying west would result in reaching the mainland.
- It was determined that Lt. Taylor was not at fault because his compasses had stopped working.
- Reconstruction of the incident determined that the islands visible to Lt. Taylor were probably the Bahamas, far northeast of the Florida Keys, and that Flight 19 was exactly where it should have been. The board of investigation found that because of Lt. Taylors's belief that he was on a base course toward Florida, he guided the flight farther northeast and out to sea. It was and is general knowledge at NAS Fort Lauderdale, that if a pilot becomes lost in the area, they are to fly a heading of 270° west or towards the sunset if their compass has failed. By the time Flight 19 had turned west, they were most likely far out to sea and had already passed their aircraft's fuel endurance.
- Flight 19 ran out of fuel and may have crashed into the ocean somewhere north of Abaco Island and east of Florida
- The 14 pilots and 5 aircraft were never recovered, and a final determination of the cause was never discovered.
- The Martin PBM Mariner flying boat launched from Naval Air Station Banana River to search for Flight 19 was also lost with all 13 crew members perishing. The PBM aircraft had a history of accumulating flammable gasoline vapors in its bilges. Professional crash investigators determined that the fate of the PBM was most likely a mid-air explosion.

UFO reports in this area are numerous with witness sightings coming from the shores of islands, countries, military ships, fishing vessels, cruise ships, and aircraft. Many of these sighting reports will be addressed in Chapter 7: Reports from the Caribbean, Chapter 8: Military, Cargo and Fishing Vessels, Chapter 9: Ocean Sightings Logged by Ships 1948 Thru 1968 and Chapter 10: Underwater Alien Bases.

Dragon's Triangle, Japan-Philippine Sea

The Pacific has a mysterious triangle area known as the Dragon's Triangle, but also known to some as the Devil's Sea. This area of water is just off the coast of Japan, and has been linked to numerous ship disappearances, UFO sightings and electromagnetic anomalies. So many Japanese military vessels vanished in this Triangle in the 1950s, that researchers were sent in to investigate. Unexplainably, these researchers and investigators vanished as well amplifying the mystery. It is documented that disappearances on this triangle section of ocean go back as far as the Mongol Empire.

Japanese legends tell tales of Kublai Khan, the fifth Great Khan of the Mongol Empire and the grandson of Genghis Khan, attempting to conquer Japan in 1274 and 1281 AD. Both attempts failed after losing his vessels and 40,000 crew members within the triangular area, due to typhoons. The Japanese believed that God had sent the typhoons to save them from their enemies. Divers and marine archaeologists have verified at least part of this legend having found the underwater wreckage of the Mongol fleets from the time and region.

In the period between the 1940s and 1950s, numerous fishing vessels and over five military vessels disappeared in the sea, in an area that lies between Miyake Island and Iwo Jima. In 1952, Japanese officials sent research ship Kaio Maru No.5 to investigate the cause of the missing vessels in the area known as the "Dragon's Triangle". Tragically, the Kaio Maru No.5 with 31 crew members, met the same tragic fate. Although the wreckage of the Kaio Maru No.5 was recovered, the crew members were never found. Soon after this additional loss and tragedy, the Japanese government declared the area known as the "Devil's Triangle" "dangerous for marine voyaging and transporting goods".

The term "Dragon in the Devil Sea's" originated from the Chinese fable telling of dragons living below the water surface. These dragons, according to a legend originating well before the AD period – 1000 BC era., lived under the sea, and attacked vessels to satisfy her hunger by means of those on-board the passing vessels. The Japanese name "Ma-No Umi," means "The Sea of the Devil", and was passed down from ancient Japanese countrymen telling tales of "paranormal phenomena" in the sea.

The area of the "Devil's Triangle" along with several other mysterious triangle areas found across the world are being studied by those in science as "Vile Vortices" areas with currents resulting in electromagnetic disturbances such as the well-known "Bermuda Triangle" It has also been suggested that subsea volcanoes erupting in the area causing "rouged waves" may be an explanation for the disappearance of missing vessels. Still, other theories are of methane hydrates on the seabed. When methane hydrates gas or methane clathrates explodes, bubbles are formed on the surface of the water, and ice-like deposits separate from the bottom of the ocean at the time of the explosion. This process can interrupt buoyancy destroying a passing vessel without leaving a trace.

Several books have been published documenting paranormal activity in the area of the "Devil's Triangle" including "The Dragon's Triangle", by author Charles Berlitz, who speaks of the unexplained loss of five Japanese military vessels and their crew.

As reported on Fox News September 3, 2009

The wife of Japan's prime minister, Yukio Hatoyama, claimed in a book called "Most Bizarre Things I've Encountered" that she flew in a UFO to Venus in the 1970s, where she made fast friends with the natives. "While my body was sleeping, I think my spirit flew on a triangular-shaped UFO to Venus," wrote Miyuki Hatoyama. "It was an extremely beautiful place and was very green."

April 19, 1957, 31° 15' N - 143° 30' E Kitsukawa Maru

Crew members aboard the Kitsukawa Maru, a Japanese fishing boat, spotted two metallic silvery objects descending from the sky into the sea. The objects were estimated to be ten meters long with no wings. As the objects entered the water, they created a violent turbulence.

MUFON Case Files Japan

As reported by witnesses

110318 2020-07-22 12:44AM Juku: UFO flash in the sky

107110 2018-12-25 6:45PM Kamogawa: Three light trails captured on camera

107083 2015-07-17 11:45PM Tokyo: Blue flash and small white flash seen on video

107081 2019-06-25 11:17PM Shirogane: seen near bottom left corner of video, two lights showing zigzag motions, keeping the same distance between them

#107074 2014-06-24 10:26PM Shirogane: You can see a bright object soaring up from the bottom right corner of the video

#105201 2019-12-25 5:48AM Onjuku: High speed light trail (two)

#102590 2004-07-03 8:00PM Kumamotoken: White light descended to me

#95723 2018-10-20 12:00AM Numazu: Many Lights forming

#94308 2018-08-21 12:00AM Seen from commercial airliner: 75-100 red orbs, lines of yellow and green orb like glow, stationary, extremely bright

#71248 2014-09-16 4:46PM Tokyo: Hiroshi Disc UFO images caught by my "Automatic UFO Observation Recording System"

#61097 2014-10-22 5:30PM Osaka: We saw this white yellowish light/craft hovering in same spot for 15 minutes above clouds

#54433 2014-03-03 9:20PM Tachikawa: Multiple sightings for over a year; 4 to 5 lights at a time, moving around, toward, and away from each other

#41060 2012-07-22 8:44PM Beppu, Kyushu: Bright light for 5 seconds

#44586 2012-11-27 5:25PM Hitachinaka: On 3 different Tuesdays, same event same time same object

#43751 2012-11-04 3:20PM Tokyo: Orb sighted over Tokyo

#31149 2011-08-25 8:30PM Tokyo: Family and friends witnessed object. "We were at home, window was open, we noticed red/blue/green/white flashing colors not moving. The lights were hovering at first then making some small circular-like movements. I called the family when I first noticed it was flashing colors and not moving. It wasn't a plane or helicopter. There are many of those around here. No fear. Interesting. It finally descended slowly in the northern direction.

#27579 2011-01-28 7:10PM Ebina: Bright as a car headlight, brighter than a star, looked like a star or helicopter

#26523 2009-03-15 12:00AM Tokyo: 3 "disc" shaped UFOs flying in triangular formation

#16172 2008-07-01 1:58PM Exact location not submitted: Photo of UFO

#6581 1945-00-00 12:00AM During WWII Combat: "My father saw creatures observing combat during a battle in the Pacific during WWII. I do not know which battle, he served in the Pacific in 1942 & 1945. He told of seeing UFOs before and after the war"

Michigan Triangle, Lake Michigan

The Michigan Triangle area has been linked to aerial UFO sightings and alleged disappearances of aircraft, boats, and ships for decades. The mysteries began back in the 19th Century, with boats frequently going missing on the lake. Historian and Mariner Mark Thompson estimated the total number of wrecks is likely greater than 25,000.

Disappearance of The Thomas Hume

The Thomas Hume was a schooner built in Manitowoc, Wisconsin, in 1870. It was christened as H.C. Albrecht, in honor of its first owner, Captain Harry Albrecht. In 1876, the ship was sold to a Captain Welch from Chicago, but the following year was bought by Charles Hackley, a lumber baron who owned the Hackley-Hume Lumber Mill on Muskegon Lake. The ship was then renamed the Thomas Hume in 1883, after Charles Hackley's business partner, and refitted so that it could safely transport lumber from Muskegon to Chicago by way of Lake Michigan. The framing of the Thomas Hume was strengthened, a new deck was built, and a third mast added to it so it could began doing its job of transporting lumber across Lake Michigan. On May 21, 1891, the Thomas Hume was last seen leaving Chicago for Muskegon with a crew of seven after unloading a cargo of lumber. Sailing along side of the Rouse-Simmons, a nearby storm raged on the lake. The captain of the Rouse-Simmons decided that it was too dangerous to continue the journey and made a U-turn back

Figure 15 Thomas Hune wreck Photo: Robert Underhill MSRA

towards port. The Thomas Hume continued on and was never seen again. When the Thomas Hume did not make it back to Muskegon, Charles Hackley and Thomas Hume dispatched mariner, Captain Seth Lee, to search for the ship but Lee couldn't find a single piece of wreckage and the seven crew members were presumed dead. Hackley and Hume continued the search offering a large reward to anyone who could providing reliable information as to the whereabouts of the missing ship, but the reward was never claimed.

The mysterious disappearance of the Thomas Hume led to many speculations including accusing the crew of stealing the ship by repainting it and turning it into another ship. Many others believed the ship was a victim of some type of paranormal phenomena and that it had sailed unknowingly into the so-called Michigan Triangle. In 2006 the mystery was partially solved with the discovery of the Thomas Hume by Taras Lysenko, a diver with the A&T Recovery Team who came across an almost completely intact vessel at the bottom of the southern portion of Lake Michigan. The ship was remarkably preserved and undisturbed because of where it sank, which was in a very deep part of the lake. A&T Recovery turned over the wreck to the Michigan Shipwreck Research Associates

(MSRA) and a respected team of Chicago underwater explorers. Various tools, coins, clothing, and jewelry were found on the wreck, dating back 115 years. The MSRA and archaeologists from the Lakeshore Museum Center, reached the conclusion that the wreck was indeed the Thomas Hume and to this day are still trying to piece together the final hours of the ship and its seven crew members.

Port Arthur, Ontario November 23, 1918

Three minesweepers the Sebastopol, the Inkerman and the Cerisoles left port together but two of them were never seen again. The weather was calm on day they left port but within 24 hours a winter storm accompanied by thick snow caught the ships off guard.
The Sebastopol lost sight of the Inkerman and Cerisoles in the heavy seas. It was thought that mechanical troubles most likely would leave them drifting onto the Slate Islands but no sign of them and no debris was never found. All hope was abandoned for the two lost minesweepers and the 76 men sailing on them.

Case of the Rosa Belle

In 1921, eleven people inside the ship, who were all members of the Benton Harbor House of David, disappeared. The ship was found overturned and floating in Lake Michigan.
It appeared that the ship had been damaged in a collision, but no other ship had reported an accident and no other remains had been found.

(By Associated Press)
Kenosha -- The steamer Cumberland has the schooner Rosa Belle of Benton Harbor, which capsized off Milwaukee, in tow two miles off Kenosha this morning and it is expected that it will be towed into Milwaukee. No reports have been received here as to the fate of the crew of the vessel.

(By Associated Press)
Racine, Wis. -- Captain ERHARDT GEISE and his crew of nine men are believed to have perished when the Rosa Belle capsized 35 miles northeast of this city.
The crew of the Ann Arbor car ferry No. 4 discovered the vessel, bottom up, but the crew was missing. The schooner cleared from High Island last Friday with a load of lumber and wood. At first it was believed that the boat had sprung a leak and filled and then capsized, but later information indicates that the ship was rammed by one of the large iron ore freight steamers and that perhaps the crew was taken on board the larger boat.

Capt. Olander, of the local coast guards, when interviewed gave it as his opinion that the crew was drowned. He said that when the wreck was reported the steamer Cumberland, with Capt. Alexander, of the U.S. marines, in charge, started out in search of the capsized vessel and found it 30 to 40 miles out. A line was attached, and the steamer started for Racine with the wreck. When about 22 miles north of here a terrific northeaster swept the lake, the tow line parted and the Cumberland ran for this port, arriving here about 3 o'clock Tuesday morning. As soon as the storm subsides the steamer, accompanied by the coast guards, will go out and tow the wrecked vessel to this port. There is a possibility, however, that the ship may be completely torn to pieces.

(By Associated Press)
Manitowoc -- That the two-masted schooner Rosa Belle was run down by a large freight steamer during the heavy fog of Saturday night, the crew picked up by the steamer and carried along on the trip, is the opinion expressed by the officers of the Ann Arbor No. 1 ferry, which arrived in port Tuesday morning from Frankfort. The ferry is the one which sighted the wreckage of the Rosa Belle out of

Milwaukee Sunday morning and reported their find to the Milwaukee offices.

According to Captain Fredrickson and Mates Axel Fredrickson and O. B. Olson, the latter, a Manitowoc man, the Ann Arbor, which has been at Milwaukee for a number of days left that port early Sunday morning and about three and a half hours out sighted a lantern screen and then a hatch floating in the lake before they discovered the Rosa Belle floating bottom side up, her stern torn out of her, the spars, gaffs and rigging on the lee side of the boat with the stern among the other wreckage. The ferry made a trip completely around the wreck but found nothing to indicate what became of the crew or the cause of the accident. The life yawl was missing.

Northwest Orient Airlines Flight 2501

Figure 16 NW DC-4 Photo: Michigan shopwrecks.org

A Northwest Orient Airlines DC-4 Pro-liner flying a daily transcontinental service from New York City to Seattle disappeared the night of June 23, 1950. Fifty-five passengers and three crew members were killed which made this crash the deadliest commercial airliner accident in American history. The DC-4 was approximately 3,500 feet over Lake Michigan, and 18 miles NNW of Benton Harbor, Michigan. It vanished from radar screens after requesting descent to 2,500 feet. A widespread search including sonar and dragging the bottom of Lake Michigan with trawlers was conducted and carried on for days. Light debris was found floating, but divers were unable to locate the plane's wreckage. DC-4 Pro-liner

Otis Redding Dies in Plane Crash
December 10, 1967

Figure 17 Wisconsin State Journal

Just 4 miles from Madison airport On December 10, 1967, a plane carrying Otis Redding and other members of his band, The Bar-Kays, plunged into Lake Monona. They were on their way to a show, when their twin-engine aircraft crashed in the lake, miles from the airport. Redding, 26, a soul singer and songwriter, along with six others, died in the crash including 17-year-old valet Matthew Kelly and pilot Richard Fraser, 26. Witnesses claimed the plane which had just radioed in for final descent, unexplainably plunged straight down into the cold waters of Lake Monona having just passed over Lake Michigan. The plane was headed to Madison, Wisconsin where Redding was scheduled to perform next. There was no distress call from the pilot. Odis Redding was famous for the song "Dock of the Bay."

Figure 18 SS Edmund Fitzgerald Photo: Robert Campbell

The SS Edmund Fitzgerald
November 10, 1975

The sinking of the SS Edmund Fitzgerald was the largest and most well-known shipwreck in all five of the Great Lakes. The Edmond Fitzgerald departed Lake Superior port at approximately 14:30 and was joined by the Arthur M. Anderson which had departed Two Harbors, Minnesota. The two ships captains, McSorley of the Fitzgerald and Cooper of the Anderson were in radio contact a mere 10 to 15 miles apart. Both captains were aware of a storm brewing so decided to take the northerly course across Lake Superior, protecting them by the highlands on the Canadian shore, then setting course southeast to the shelter of Whitefish Point.

As the weather worsened, Gale warnings were eventually issued at 19:00 on November 9th and upgraded to storm warnings early in the morning of November 10th. Winds were gusting to 50 knots with waves at 12-to-16-feet. By 17:20, winds had reached 58 knots with gusts to 70 knots, with 18-to-25-foot pounding waves. At approximately 18:55, Captain Cooper of the Anderson stated that he "felt a "bump", felt the ship lurch, and then

turned to see a monstrous wave engulfing their entire vessel from astern." This forced the bow of the Anderson violently down into the sea. Monstrous waves continued to pound both ships which kept in constant radio contact to assure each other's safety.

Clark, the first mate of the Anderson, spoke to the Fitzgerald one last time, at 19:10: "Fitzgerald, this is the Anderson. Have you checked down?"
"Yes, we have."
"Fitzgerald, we are about 10 miles behind you, and gaining about 1 1/2 miles per hour. Fitzgerald, there is a target 19 miles ahead of us. So, the target would be 9 miles on ahead of you."
"Well," answered Captain McSorley, "Am I going to clear?"
"Yes, he is going to pass to the west of you."
"Well, fine."
"By the way, Fitzgerald, how are you making out with your problems?" asked Clark.
"We are holding our own."
"Okay, fine, I'll be talking to you later." Clark signed off.
Due to the storm, it was difficult for the Anderson to keep a radar signal, and at approximately 19:15, radar contact with the Fitzgerald was lost and did not reappear. First mate Clark called the Fitzgerald repeatedly, but there was no answer.

Captain Cooper radioed the Coast Guard at 20:00 expressing his concern for the safety of the Fitzgerald. The Anderson reached Whitefish Bay, Lake Superior safely, but at 21:00 at the request of the Coast Guard, set sail again in search of the Fitzgerald. Several other ships and aircraft were involved in the search for the Fitzgerald, but as feared, she was gone. The SS Edmond Fitzgerald was lost in the storm on November 10, 1975, near Whitefish Point, Lake Superior, there were no survivors.

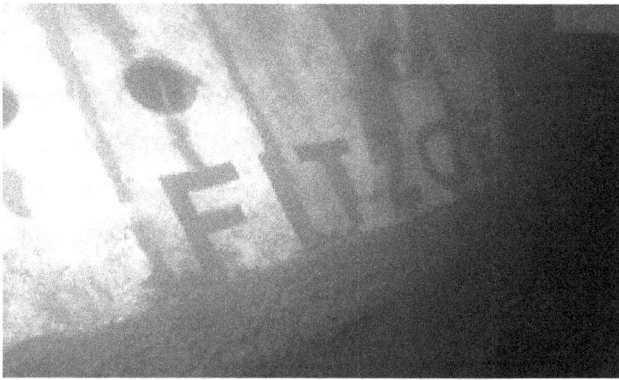

Figure 19 SS Edmund Fitzgerald Photo courtesy of Robert Campbell

The unexplained activity in the Michigan Triangle area continues and is covered in Chapter 6: The Great Lakes Region

Bennington Triangle, Vermont

Vermont is another geographical area that has become notorious for mysterious activity. The Bennington Triangle includes the landmarks of Glastenbury Mountain and several known ghost towns which have been known for a number of unexplained disappearances over the years. These disappearances began in 1945, when a local man named Middie Rivers was leading a group of hikers near the mountain. He walked ahead of the group and vanished into thin air. The following year, teenager Paula Welden was hiking in the same area when she also disappeared, triggering a media frenzy and massive hunt. Middie Rivers and Paula Welden were ever found. A few years later, a young boy named Paul Jepson was told to wait in his mother's car while she visited a site near the mountain. When she returned, her vehicle was empty. A huge search was launched, with hundreds of volunteers scouring the area but Paul Jepson was never found. Paul Jepson and Middie Rivers were both wearing red clothes when they vanished, which led to the superstition that it was bad luck to wear red and visit Glastenbury Mountain. Over the years many others have vanished without a trace.

numerous UFO sightings have also been reported in this triangle area.

Bridgewater Triangle, Massachusetts

The Bridgewater Triangle covers some of the most rugged and mysterious terrain in the state of Massachusetts. That are vast wetlands and wooded areas large and dense enough to get easily lost in. In this area, one of the very first UFO sightings in the U.S. was documented back in the year 1760, a "sphere of fire" so brilliant that it was seen across several towns. In 1908, a pair of undertakers who were travelling on a carriage, reported seeing a flying object they described as "an unusually strong lantern". This was seen by multiple witnesses who corroborated their story. Many believed the object was a hot air balloon, but the undertakers did not agree. One writing, "I claim that a hot air balloon could not move in a circle or perpendicular as this one did." Since then, several unexplained and mysterious reports have come out of the Bridgewater Triangle, including sightings of Bigfoot, Moth-man like creatures, and animal mutilations.

Matlock Triangle, Great Britain

The Matlock Triangle is a zone in Great Britain located in the Derbyshire Dales known to some as the UFO capital of the world. Numerous sightings of odd, glowing objects hurtling over the Dales have been reported. These are reported as coming from cigar-shaped aircraft and one described as being shaped "like a bowler hat". The most widely reported sightings in the area were made by Sharon Rowlands, a local woman who considered herself a "complete and utter disbeliever" until she saw a giant multicolored disc hovering over her village. Ms. Rowlands caught the sighting on her camcorder, which she reportedly sold to a US TV company for a large amount of money. Multiple sightings of

unexplained craft have been reported over this
area of water.

Chapter 5

The Alaska Triangle

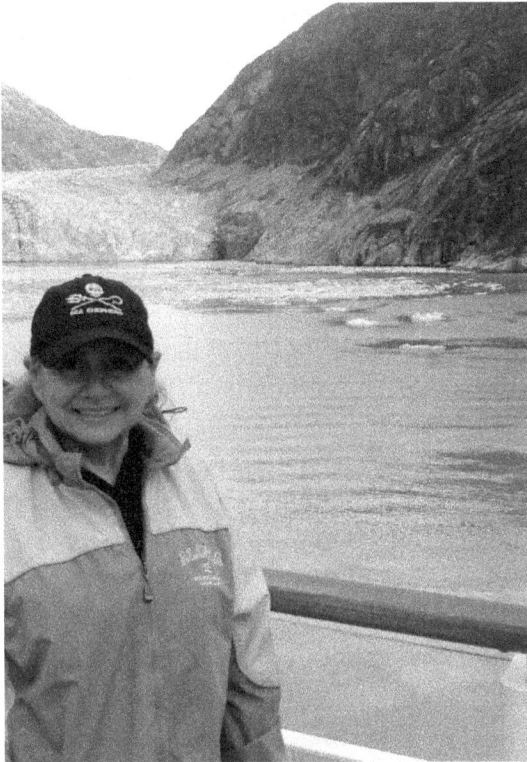

Figure 20 Debbie Ziegelmeyer in Alaska
Photo: Wayne Ziegelmeyer

The Alaska Triangle covers an area so massive an entire chapter is necessary to do it justice. This massive region is an area of wilderness connecting the three city points of Barrow at the tip, Juneau to the bottom west, and Anchorage on the bottom east. This area also includes the northern region of the Barrow Mountain range. This area within the triangle has been researched and talked about for decades due to multiple mysterious disappearances, unidentified creatures, and UFO sightings both above and coming up from underwater.

Within this zone there are vast areas of largely unexplored wilderness, forests, mountain peaks, and desolate, barren tundra. This region has an unusually high number of people, both tourists and locals, who go missing every year without a trace as if they have vanished from the face of the earth. The Alaska Triangle has a long history of planes disappearing or crashing and ships small and large, sinking, never to be found. Since 1988, a staggering 16,000 thousand people have vanished in the "Alaska Triangle," never to be heard from again. The rate of people reported missing in Alaska is almost twice the national average. While many cases involve runaways or people who return home, Alaska also has the highest percentage of missing people who are never found averaging out at a rate of 4 in every 1,000 people.

October 1972

A Cessna 310 carrying House Majority Leader Hale Boggs and Representative Nick Begich, along with an aide, Russell Brown, and their bush pilot Don Jonz, mysteriously vanished in the region while on their way from Anchorage to Juneau. The search lasted for 39 days involving over 400 aircraft including an advanced Air Force SR-71, and dozens of boats, which included twelve from the Coast Guard. Even with all that effort, no evidence of what happened to the three men, or the plane was ever found, and the men were declared dead. To this day, no trace of the men, plane, or scrap of metal, has ever been found.

November 16, 1986

Figure 15 Photo taken by Debbie Ziegelmeyer over Alaska

A UFO was sighted by the crew of a Japanese jumbo freighter aircraft which had been detected on radar by Air Route Traffic Control Center in Anchorage, Alaska. The flight route of the Boeing 747 was from Paris to Anchorage, Alaska and then to Tokyo with a stop at Reykjavik, Iceland. The airplane with a crew of three (Captain Kenju Terauchi, First Officer Takanori Tamefuji and Flight Engineer Yoshio Tsukuda), reported its position to the flight control center at Edmonton, Alberta, and continued over northern Canada, flying at an altitude of 39,000 feet. They began making final preparations before descending to Anchorage Airport when suddenly Captain Terauchi and his crew spotted unusual white and yellow lights ahead, below, and to the left of their airplane. At first, they thought these were originating from military aircraft but quickly realized they were not. First officer Takanori Tamefuji made radio contact with

Anchorage air traffic control and asked if there were other aircraft in the vicinity. Both Anchorage and a nearby military radar station informed them that they were picking up weak signals from the 747's vicinity but were not sure what they were seeing. The crew of the 747 observed one huge lighted object and two small lights which were insight on their radar for more than a half-hour. They were able to see this unidentified craft as they flew 550 km (350 miles) southward across Alaska from Ft. Yukon toward Anchorage Airport.

Captain Terauchi Stated the Following About the Unidentified Lights:

"They were flying parallel and then suddenly approached very close..."

The captain also caught a short glimpse of the main object's walnut-shaped silhouette and estimated it to be "two times bigger than an aircraft carrier". The flight crew of the 747 saw and tracked three unidentified objects. Captain Terauchi later wrote in a statement for the Federal Aviation Administration that after several minutes, the lights suddenly darted in front of the 747, "shooting off lights" that lit the cockpit with a warm glow. Sitting in the left-hand seat, he saw the lights in front of and below the airplane, when they began to move erratically, "like two bear cubs playing with each other." As the Boeing 747 was passing over Eielson Air Force Base, near Fairbanks, Captain Terauchi noticed something very large and unusual behind his airplane. It was the dark silhouette of a gigantic "mothership" larger than two aircraft carriers which was clearly visible in his field of view. Captain Terauchi asked air traffic control for permission to take his airplane around in a complete circle and then descend to 31,000 feet. As he made this maneuver and several others, his plane was followed by the objects exactly through all the maneuvers and during several turns for thirty-two minutes before it finally vanished. Captain Terauchi also reported that the UFOs moved very quickly and were able to stop suddenly, and that the light from the gigantic "mothership" was so bright, it lit the airplane's cockpit, and he could feel the light's heat on his face. Later, the Japanese pilot admitted the entire crew had been watching the UFOs for about six minutes before notifying anyone on the ground.

Reported by FAA Agent Ronald E. Mickle After Interviewing the JAL Crew
"Captain Terauchi stated he first sighted (visually) the unidentified air traffic (UAT) in the vicinity of Potat intersection and the ADIZ (Air Defense Identification Zone). The aircraft he was piloting (B747) was at flight level 390, airspeed 0.84 Mach... the UAT was in front of his aircraft at a distance of approximately seven to eight nautical miles for approximately 12 minutes... the distance was indicated by the onboard Bendix color radar. Captain Terauchi stated that while he had a visual on the UAT, he spotted yellow, amber, and green lights, and a rotating beacon, but no red lights... there were two distinct sets of lights but appeared to be joined together (as fixed to one object). Captain Terauchi stated that during the visual sighting, the lights of the UAT changed from a horizontal position to a vertical position and had positioned itself from in front of the B747 to the port side. The UAT stayed on the port side for approximately 35 minutes. Captain Terauchi stated there was static during VHF communication with Anchorage and there was an erratic movement with lights of the UAT during the visual contact. FAA, ATC (Air Traffic Control) had indicated to him the presence of a primary target in addition to his aircraft."
After the FAA investigated the incident and found the crew to be "normal, professional and rational", the US Federal Aviation Administration authorities admitted that the objects were tracked on radar and were not registered on the radar tapes.

Witness Sighting Accounts
Source of the Following Alaskan Triangle UFO Sighting Reports
MUFON (Mutual UFO Network)
NUFORC (National UFO Recording Center, Peter Davenport)

Ketchikan, Alaska
Friday March 11, 2016
Shape: Sphere
Duration: Undisclosed
Source: MUFON
I was at home in Mendocino a County, California using my iPhone to view images of a Ketchikan, Alaska webcams. Ketchikan Public Utilities (KPU) operates five or six views of the docks, water, distance. views, town. Every few minutes the cameras update. Today was a nice mostly clear sunny day there. So, I captured a few images, saving them. I went back into the cams. It was then I noticed the dark orb in a south facing cam. Reviewing all five cams only the two facing south had the orb. And from two separate locales with two separate views, each confirming the orb as real and not matter on the camera lens. The object was stationary but did drift slightly. I could tell by lining up distant mountains and closeup objects. I was very busy today but kept checking in or the cams and capturing the object. After about three hours, one cam lost the orb. What was weird is on the other cam it was still there. By the time I checked again clouds had moved in. The object was gone.

Ketchikan, Alaska
Monday March16, 2015
Shape: Circle
Source: NUFORC
There were 2 glowing orange balls dropping smaller white balls, they blinked then disappeared. We saw 2 orange glowing balls in the sky, they seemed to be dropping smaller white balls. They looked like they were chasing each other, moving around in the sky. They

were blinking, then they disappeared completely.

Ketchikan, Alaska
Tuesday December 30, 2014
Shape: Star-like
Duration: 00:05:00
Source: MUFON
My husband and I were standing on our porch, looking up at the sky for northern lights. When all of a sudden, we looked out towards the end of the runway of the airport. The airport is on an island with woods surrounding it. About 1/10th of a mile from the end of the runway, back in the woods, we saw a star-like object rise above the trees and ascend into the sky. It was moving at medium speed, and we knew it wasn't a helicopter or any kind of human aircraft. The object kept going up and up until it stopped, about 500-1,000 feet in the air. When it stopped, it looked exactly like a star, and the clouds covered it and I didn't see it again. 3 minutes later an Alaska Airline jet landed. About 5 hours later, we saw the same thing, but it was more towards the middle of the runway. It had gotten cloudy, so I did not see the object stop, as the clouds were blocking my view. I have no explanation for this!

Ketchikan, Alaska
Tuesday August 19, 2014
Shape: Oval
Source: MUFON
I was on a small cruise ship touring in the Ketchikan area (the Tongass Narrows) with a group of co-workers and friends. At about 8:00 PM, I was taking some photos of the sunset with my iPhone. I took three shots of the sunset. The following evening, as I downloaded the photos to my computer, I noticed the bright light in the sky (upper left corner of all three photos at slightly different angles since the boat we were traveling was heading north or north-west as I was taking the sunset photos. I did not notice the bright light in the sky as I was taking the photos, yet the light object is clearly in the photo. Since I did not see the object directly at time of the photo, I cannot say that it was actually moving in the sky. On the photo the object or light appears oval of possibly disk shaped. I have attached 3 original digital photos and one that I enlarged and enhanced.

Ketchikan, Alaska
Thursday August 7, 2014
Shape: Oval
Duration: 2 minutes
Source: NUFORC
Daylight a few miles away, very bright, oval, silent, stationary then disappeared. About 5:30 pm on August 7, when we looked out across the water from my house, and we saw approximate one mile away a very bright solid object in the sky. Perhaps it was bright as the sun could have been reflecting off what appeared to be a very shiny metallic surface. Oval in shape. It appeared to be very slowly moving away from our direction. Only observed it for a minute or so before it just suddenly disappeared as though it went into the clouds, yet the clouds were too far away for that to be the case. It was silent. Never seen anything like that, it was definitely a solid unusual object in the sky, brilliance like never seen before. Neither of us have ever seen an UFO before, I personally feel this was a UFO

Ketchikan, Alaska
Thursday December 5, 2013
Shape: Light
Duration: 16 minutes
Source: NUFORC
Light stayed over the airport and slowly went left to right. Then it got really bright and dimmed and was gone. Sighting was over the airport island just below the moon. A friend called from his work about 2 miles from my

home. My friend said hurry go outside look at it, so I did. The light stayed over the airport then slowly went left to right. Then got really bright and dimmed and was gone. I had my friend on the phone. He could still see it. I said it went dark after I started taking pictures. He said he could still see it. He said it went up then went dark. I took pictures pretty fast. The first picture I took I got the light, second picture it's gone. The light stayed over the airport slowly went left to right, then got really bright and dimmed

Ketchikan, Alaska
Saturday September 7, 2013
Shape: Disc, Fireball
Source: MUFON
We have an expansive view over Nichols Passage in SE Alaska. My wife, and I were sitting at home watching TV on September 7th, 2013, 9:25 PM when I noticed a bright light, like a fireball to the SE dropping through the cloud layer at 4,000 feet approximately. This is a nightly vigil watching the Alaska Airlines jet on final approach to Ketchikan International Airport. Only this light was the wrong color, brighter and traveling too fast. I said to my wife, get my cell phone and take a picture of this thing, as I went outside to get a better look with binoculars. This object which looked to me to be traveling at around 500-600 mph at 4000 feet and 3-4 miles distant from my location, it made a slow left turn at Blank Point and flew west down the Tongass Narrows over the airport and disappeared. My wife was going back in the house when I said wait, there's another one coming! I watched the second one with binoculars following the same flight path that all aircraft using IFF use, (Bushnell 10x 50's). It appeared to be a glowing disk of plasma, or like churning molten metal with 6 white lights rotating around this "molten" disk in a rapid pulse pattern, extremely bright. I did not see the third UFO that my wife saw following the second UFO

because I was looking through binoculars limiting my field of vision watching the second UFO up close for approximately 10 seconds. They made no sound. Our location gets every sound bounced off the mountains by aircraft and boats and these UFOs made no sound at all. They all disappeared as the first, into thin air or into the cloud cover. I'm a professional artist and registered guide, a trained observer. I have no reference in my experience of 56 years to compare this to.

Ketchikan, Alaska
Thursday November 8, 2012
Shape: Light
Duration: 15-20 seconds
Source: NUFORC
A white light flew across the sky and then shot straight up. It faded to red and disappeared. I saw a really bright white light when I looked up at the sky, thought it was a satellite but realized that it was too close and too bright, it flew across the sky for about 10 seconds and that's when I realized it wasn't anything that I could explain, it wasn't flying straight, it moved back and forth just barely and then the light got dimmer and dimmer. It looked like it was going straight up and then as the light almost dimmed out it turned red and then disappeared.

Ketchikan, Alaska
Tuesday September 4, 2012
Shape: Circle
Source: MUFON
We all went out to dinner, but we took two different vehicles because we arrived from two different directions. After dinner I took our five-year-old and drove behind my husband (who took his mom and our two-year-old). The island we live on is very remote and the city is very small. There's only a two-lane road, no freeways, and only one airport with one major airline that comes in and out several times per day (with exception to FedEx and UPS, etc.).

After dark all float plane activity stops and there no helicopter activity either unless the coast guard gets an emergency rescue call. We were about half-way home (a six-mile trip one-way) when my daughter exclaimed "Mom! Look at that thing!" and she pointed up to the sky where this bright orangish-colored light similar to the color of your typical high-pressure sodium streetlight, high up in the sky. At first, I thought it was a light from a tall pole but then realized it was way up in the sky at cloud level, but it was not moving like a helicopter or airplane and there were no blinking lights on it. There was, however, this second source of light that was rotating from it like giant surge light. This light was coming from a very thin cloud in a relatively clear sky. My husband saw it too and he pulled over to the side of the road. I parked behind him and met him in front of his car and we both just looked up in amazement trying to figure out what this "thing" was! We both could see the revolving light and were dumbfounded by the lack of any blinking lights as seen on helicopters or airplanes, and the stationary way it moved, which was very slow and going upwards. Then, after only a minute, it turned 180 degrees away from us, and totally disappeared. The cloud that it was near was too thin for it to have completely concealed it. Furthermore, my mother-in-law saw it and was positive it was not any type of aircraft she has ever seen. When I got back in my truck, my daughter said, "Mom, that was an alien spacecraft!" We haven't talk about aliens with kids because they're so young and the subject hasn't come up yet. Then she just HAD to see my cell phone. She took it and said, "I'll show you what that was!" 45 seconds later she shows me this picture from a coloring book application I have on my phone, of a flying saucer with a beam of light emanating from it! My husband has absolutely no explanation for what that was and he's an engineer that has a logical explanation for everything! I have never witnessed anything like it, but it definitely caught all of our attention.

Ketchikan, Alaska
Saturday April 28, 2012
Shape: Sphere
Duration: 00:06:00
Source: MUFON
I arrived home at 10 pm. As I got out of my car. I heard a flight of Canadian geese flying northward. When I looked up, I saw 3 bright lights about 1000 feet above Tongass Narrows. There was a white light on top, a white light below, and a red light in the center. The lights were slowly flying along with the flock of geese. When I first saw the lights, I thought it was very out of the ordinary since small planes do not fly in our area after dark. I posted my sighting on my Facebook page. The sighting was also reported by several Facebook users. One person was able to take a photo with his iPhone and posted it on Facebook. I lost sight of the object when it and the geese passed in back of a hill. The Facebook photo shows white geese flying in formation and the bright lights of the UFO.

Ketchikan, Alaska
Tuesday September 14, 2010
Shape: Blimp, Chevron, Circle, Sphere, Square, Rectangular, Triangle
Source: MUFON
As a follow-up to the report, I submitted last night: we continued to observe the lights and I e-mailed the following update last night to MUFON. Mon, 9/13: The larger multi-colored object we observed is still visible although it is somewhat smaller and appears to be a little further south of where we first observed it. Since I submitted the report, we have observed multiple triangular, or chevron shaped objects appear around the larger object. The three points of the smaller chevron shaped objects appear to be smaller versions of the larger multi-colored object and they

appear to pulsate and dim more so than the larger object which remains fairly static in brightness. From 9:45 - 9:54 PM Ketchikan, AK time two chevron shaped objects phased into view surrounding the larger object and "pointing" at it. Two triangular shapes were at roughly 7 o'clock and 4 o'clock in relation to the larger object. At 10:03 we observed a third chevron appear at about 11 o'clock from the larger object. The third chevron was also "pointing" at the larger object. It is now 10:51 and the larger object is still visible. I discontinued observing shortly after 11 PM on 9/13. Tonight, 9/14, we returned home at about 8:20 PM Ketchikan time and observed the same multi-colored formations. The largest one is situated more south-southeast than last night, but it is in the same general area, although it appears a little more distant and smaller. There are a number of smaller multi-colored objects as well. We haven't seen the triangles form tonight. In addition, at about 8:30-8:35 PM we noticed a brilliant white object appear to the west of Deer Mountain. When we arrived home at about 8:20 PM tonight that object was not in the sky. My sister reminded me that she mentioned the brilliant white object last night, but we were so wrapped up in observing the multi-colored objects that we didn't pay much attention to the white object. Upon closer inspection tonight, the white object has a large, brilliant light on the north side, and it appears to have mass between the north end and what appear to be small windows further back towards the south end of it. Overall, the shape appears to be flatter and more elongated than the spherical multi-colored objects. Both objects are at roughly the same altitude and appear to be stationary. A group of four of our neighbors observed these objects with us tonight. One of our neighbors has lived in the area for a number of years and she says she's never seen anything like this. Everyone agreed that these appear to be "UFO's" Again tonight, the sky is

clear, and visibility is excellent. There appears to be less erratic movement than we observed last night. The objects remain clearly visible by the naked eye as I type this at 10:10 Ketchikan, AK time.

Ketchikan, Alaska

Monday September 13, 2010
Shape: Sphere
Source: MUFON

We were standing on the balcony of our house in Ketchikan, Alaska. It's a clear night and my sister and I noticed an odd formation of multi-colored lights hovering over the southeast area of town. We noticed it at least 45 minutes ago and we've been observing it for about 45 minutes with binoculars. It's hard to determine the exact shape but it appears to be somewhat spherical, not making any apparent forward or backward progress, and it's maintaining a fairly steady altitude. We haven't seen anything like this before. Sometimes it appears steady then it zigzags erratically. The colors range from red, blue, yellow, white, purple, green. We haven't lost sight of the object. It appears to be a little bit smaller right now but it's still there and visible with the naked eye. Has anyone else reported this? Does anyone have any idea what it might be? We come from a family of pilots, and this is definitely odd.

Ketchikan, Alaska

Friday January 30, 2004
Shape: Light
Duration: 4 seconds
Source: NUFORC

Explosive light over Alaska skies. I don't have pictures but am in search of someone who can tell me what's going on in the skies over Southeast Alaska. Not only have we, as well as many others in town, seen numerous times, unexplained bright lights in the sky that have flashing red and blue and move like a UFO, but tonight topped them all. While standing on the patio, for about 4 seconds I saw an extremely bright object flying in a downward motion with lights that looked like explosions that were so bright, they shone through the clouds. I've seen falling stars and meteors; this was definitely a UFO. I remember I saw a documentary once that showed pictures from the atmosphere of spaceships flying around Earth and then what looked like gunfire from Earth towards them, like we were shooting at them. This is exactly what it looked like. Just wondering who can tell me what's going on over our Alaska skies.

Ketchikan (area), Alaska
Thursday 25. May 2000
Shape: Oval
Source: NUFORC
30 light oval shape circles. I was standing on board a cruise ship watching the sun set when there seem to be a circle of lights not moving, just standing still in the sky. I took a photo, and the image is clearly shown on film. The image is oval, and it clearly shows about 30 lights in an oval shape with one of to the side. I had the photo lab examine it to see if it was dust specks or some other form of interference, but it seems to be a UFO. NUFORC note: We have received a photographic print from the witness, and we share his interest in this event. It is impressive, we feel. We have no explanation for what the objects could be. If anyone else witnessed the lights over Ketchikan on this date, we would like them to call, or submit a report over our website. PD)

Ketchikan, Alaska
Wednesday August 11, 1999
Shape: Circle
Duration: 7 seconds
Source: NUFORC
I saw a bright object in an area where there is or never has been anything. I am 60 years old and very familiar with aircraft. 12 years with Boeing Company, my brother-in-law is a pilot who owns 20 airplanes. I write children's

books and spend lots of time outdoors in southeast Alaska. I know how remote we are and up here we have virtually no night flying by anything but commercial aircraft. At 11:07 pm on 8/11/99, I was sitting on my bed looking north over the city of Ketchikan. I was watching a star that is generally the first star that I can see. Above the star was a bright light and I mean bright, appeared the object. It was round and probably as big as an index fingernail. It was moving southeast slowly, much slower than an aircraft. I picked up my binoculars and looked at it, and it was so bright it dazzled like a diamond and gave off white-blue light. I watched the object for about 7 seconds and then went out on my deck to further observe and it was gone. It didn't reappear. 2 days later 8/13/99 at approximately 8:30 pm in the same area, I saw a round object coming towards me. It was silver with a dark center band. Visual time maybe 3 or 4 seconds. It turned to the west and disappeared.

Ketchikan, Alaska
Saturday 05. August 5, 1978
Shape: Rectangle
Duration: 3 minutes
Source: NUFORC
Saw an aircraft carrier shaped UFO During the summer of 1978, while I was fishing near Ketchikan. I looked across the ocean at the island in the distance. I was shocked when I saw a huge, gray UFO that looked similar to an aircraft carrier, hanging in between two mountain tops. I did not notice any lights on the craft, but clearly saw a soft shiny, metallic hue surrounding it. As I sat in my boat with chills running down my neck, it hung between the mountains for several minutes. Then, in a blink of an eye (literally), it simply disappeared from sight without making a sound. Having spent my entire life in Alaska, I have witnessed many strange sights, but not many that compared to this UFO. Since this sighting, I

have witnessed strange lights hovering in the sky and then moving quickly out of range, but I have never again had the pleasure of seeing this particular craft.

(Davenport, n.d.) (National UFO Reporting Center) Witness Sighting Reports

Anchorage, Alaska
3/10/2019 20:15
Shape: Light
Duration:10 minutes
They hovered in the same spot for 2 minutes then lined up next to each other. The middle one disappeared but the other two remained in the same spot. Then the middle one appeared next to the other light on the right-hand side. Then the other light that was on the bottom disappeared and the other two now remained still in the same spot.

Eagle River, Alaska
3/1/2019 13:28
Shape: Cylinder
Duration:5
Driving down Eagle River Road, I looked into sky and notice several oval shaped UFOs. I stopped to get a better look as they slowly disappeared.

Kenai, Alaska
2/13/2019 07:47
Shape: Cylinder
Duration:1 minute
Approximately 7:47 am, we all noticed one bright light and two smaller dimmer lights directly in line and behind it coming down at an extremely sharp angle like it was a plane coming in for a landing near the airport. There was no sound and the lights appeared to turn away from us slightly during descent and then simply and instantly disappeared. The lights

appeared almost squarish shape on a long thin object.

Wasilla, Alaska
1/20/2019 19:45
Shape: Light
Duration: 1-2 minutes
Looked exactly like a star, but it traveled across the sky.
I was watching the lunar eclipse at totality, and I saw something out of the corner of my left eye. I turned to see a moving star within the Ursa Major constellation (Big Dipper). The light looked the same size, shape, and color as the other stars in the constellation, but it moved. The Big Dipper was facing mostly vertically in the sky with the dipper part upwards and the handle downwards, slightly tilted to the right. The UFO moved from the upper right corner down to the lower left corner of the four stars that make the dipper part of the constellation. For a brief moment it blended in perfectly with that star. Then it curved to the right a bit and continued moving straight further away into the distance until I couldn't see it anymore. It moved quickly, but not fast. At first, I thought it was a shooting star, but it wasn't fast enough. Then I thought it was an airplane, but it was too high in the sky and the lights didn't flash. There also weren't any red lights visible. Since it was moving away from me, it couldn't have been the headlight on a plane. It could have been a satellite, but it was "twinkling" the way stars do, and it changed directions.

Anchorage, Alaska
11/30/2018 09:00
Shape: Light
Duration:30 minutes
Green light in sky after November 30th earthquake near Anchorage, Alaska.
Heard on Radio Station at approx. 9am 11-30-18 after 7.0 Magnitude Earthquake near Anchorage, Alaska. 2 people called in Station saying they saw a green light in sky after our

major earthquake. I never heard anything further.

Anchorage, Alaska
11/17/2018 19:15
Shape: Light
Duration: 15 minutes
I live in Wasilla and was hearing a weird rumble for about 10 minutes. Not thunder. Not jets flying over. Went out and looked over Cook Inlet and saw bright white lights in a stack. Not moving. One would blink out then another until there were only 2 then more would blink back on. The most I counted were 5, all lined up in a stack. They were not moving and there were no other lights other than very bright white. No green or red. At one point they all went out. Then one blinked back on and lit up the sky. Then another blinked on. I watched while the weird rumble continued. After about 10 minutes, all blinked out and the rumble stopped.

Wasilla, Alaska
11/14/2018 08:30
Shape: Circle
Duration:10 minutes
Long black fast forming cloud turning white, then a silver disc appeared.
Long black cloud forming across the sky faster than normal, turned to skinny white cloud then round disc appeared and went in an easterly direction, but not very fast.

Soldotna, Alaska
10/29/2018 20:15
Shape: Fireball
Duration:3 minutes
Driving home from work and spotted 2 orange lights above the tree line. I made a turn and then saw 5 orange lights in a straight line seemingly descending slowly in unison.
They eventually fell behind the tree line so I couldn't see them anymore. All 5 were a

steady orange light, not bright enough to be flairs and always in a perfectly straight line.

Soldotna, Alaska
10/29/2018 20:15
Shape: Light
Duration:5 minutes
Orbs orange red lights.
Observed 3 objects in night sky just over horizon. There were 3 groups of at least 3 orbs of orange red lights. Appeared to be hovering. Each appeared to flicker out one at a time.

Ketchikan, Alaska
9/29/2018 20:07
Shape: Circle
Duration:1 minute
We just got home from shopping and it's a clear night, looked up at the stars while getting out of the car and noticed a bright light moving across the sky from north to south. The light reminded me a LED light, it was very white and bright, no noise, no blinking lights. It was moving at a slow steady speed. We watched it for a minute and then the light turned to a dim red light, the red dim light got smaller and then it just disappeared.

Wasilla, Alaska
6/27/2018 14:00
Shape: Sphere
Duration: unknown
I see Chemtrails often but have discovered Orbs around my house. I can't see them, except in Photos. After getting a Chemtrail being sprayed (05/15/2018) on video, I began taking photos of planes flying over. When I looked at the photos, I saw shiny silver Orbs in the photos. As I took more picture's I got these Orbs in trees and bushes all around my house. As of yesterday, I am still capturing Orb's in photos.

Anchorage, Alaska
5/8/2018 18:00
Shape: Cigar
Duration:20 minutes
Horizontal white cigar shape UFO in Anchorage, AK. Me, three friends and my friend's mom were in the car on the way to get food. We saw a shiny weird object in the sky, and it wasn't really moving much, and there was only one of them at the time. It was a horizontal cigar shape, and it was white/silver. It wasn't a plane, and it was too high in the sky for it to be a balloon or sign. We aren't exactly sure how long it stayed there, as we were in a moving car and eventually it was out of view. It also appeared to be pretty far away, so it could have had lights, or it could have been moving more if we observed it up closer.

Anchorage, Alaska
4/1/2018 21:30
Shape: Changing
Duration:5 plus minutes
So, standing on my porch having a smoke, and what I thought was a dark cloud over downtown Anchorage, I noticed it slowly moving but not in a cloud movement, you could see it move to see its length and then its width. It was a dark grey object. I grabbed my binoculars. When I returned to get a better look, it was farther away and when I looked from what I observed it had a "smokey trail". I have no idea what I saw.

Anchorage, Alaska
3/11/2018 18:33
Shape: Cylinder
Duration:1 minute
Metallic cylinder with no lights hovering then moved rapidly west and up into clouds.
A stationary cylinder-shaped object was seen hovering just below cloud level. It was metallic looking with no exterior lights. At first, I took it for a flock of birds, but then realized it was a single object of undetermined size. It did not

move for approximately 1 minute, it then moved rapidly laterally to the west and then rapidly up into the clouds in a motion impossible for any aircraft.

Wasilla, Alaska
3/8/2018 01:00
Shape: Cross
Duration:10 minutes
I didn't go anywhere or do anything I normally wouldn't do. I was standing at my window smoking a cigarette and as I was putting it out. I looked up and saw an air vehicle that I thought was a Cessna, that was entering my field of view from the left. It was over the back of Wasilla Lake when I saw it. It moved slowly like a plane and seemed to look like a concept vehicle at over a couple thousand feet. It had a red light you would see on the wing of a plane and an engine sound. What I couldn't focus on for a moment was when the engine quit running and I thought it would crash. It was a sound effect of a plane that stopped. It turns towards me and begins to bring the bottom of the craft upright as it's approaching. As this occurred, I put my hand up slowly over my heart and away in front of it. There was one slow turn until it stopped about 40 feet in front of me. I acted like shaking its hand and returned my hand over my heart. As I looked it over, it was floating in front of me pointed parallel to the building I was in and not at me; it had a cross reaching all the way to the sides with no unnecessary markings except for the red light. This red light was situated on the top part of the cross which represents intelligent action in the way it was on top since it turned towards me from the lake. I was a few feet from my phone, but I was stunned to move naturally, and I had my girlfriend sleeping in bed right next to me. I didn't want to wake her. After about five minutes as it seems to be off to my right like presenting me with an award or something, it starts to leave to the left, slowly and in the same position it was in

last. Professionally, it seemed it was checking on a post nearby. I have experience because I'm a veteran so perhaps maybe I was more valuable than the regular person. It was excellent and exciting to witness, and I had love in my heart, and it made sense to me. I'm not cruel. Nothing around it was moving, and it became stationary when it stopped. I didn't have the motivation to do more than witnessing this astronomical event.

Mile 35, Alaska

1/28/2018 20:49
Shape: Light
Duration:6 minutes
Highway 1 Mile 36 nine lights observed. While driving North on Alaska Highway 1 at 8:49PM, I saw what I thought was a radio tower that had red lights, abouts 2 miles in front of me. I thought to myself when did someone build a huge tower out here by the Y junction? I was confused with what I was looking at while trying to drive on the icy road. As I approached closer, I then observed that there was no tower and just 9 red lights that were aligned in a row vertically. My first impression was someone is setting Chinese flying lanterns. I then began to observe the lights were moving about, and 8 turned color to an amber orange color. I noticed another vehicle coming my direction in the opposite lane. I slowed to about 25mph and pulled towards the shoulder of the highway. When the vehicle passed, I then thought about getting my cell phone and stopping to take pictures or video. I changed my mind because I had my 2 children asleep in their car seats. So, I kept driving slowly, and kept looking at the lights which are now within 400 yards of my van. I remember looking at my dash display and it read -1 degrees outside. Visibility was unlimited and it was cold and clear out. I could see the stars out and these lights were definitely not stars. These lights were not drones. I've seen drones and it was below

zero. Who would be flying 9 drones in frigid weather anyway I thought? As I passed the lights, I noticed a red light just above the trees that began to travel upwards moving in an intelligent manner among the other amber orange lights. I kept driving until I got to a pullout at the Y junction. I got out of my van and looked at the direction I came from. Now I can only see 1 red light that is going vertically up into the night sky. I lost track of it and then got back inside my vehicle. I then called my family in Seward and described to them what I just seen. My sister was going to check Facebook to see if anyone else put it online. I counted 5 other vehicles coming and going during this period. Another black pickup was parked above me and what appeared to be viewing this also. I continued North towards Anchorage and I haven't heard of anyone else making a report of the lights that night.

Fairbanks, Alaska

1/26/2018 06:20
Shape: Circle
Duration:2 minutes
Blue glowing disk with no center giving off bright green aura like exhaust.
It was flat like a Frisbee with no center. The entire underneath was a blueish color. It had a bright green (mistaken it for the aurora at first glance) gas like aura as it hovered, then it started to move. Shortly after leaving a glowing gas cloud behind, it shot upwards which my photos show. I frantically tried to get the camera set up. Missed the glowing craft itself, but the exhaust photo corroborates my claim. This thing was massive. Several 747 jets could have fit within the circumference.

Wasilla, Alaska
1/13/2018 13:00
Shape: Formation
Duration:5 minutes
Nighttime extra activities in neighborhood. No people visible. Multiple bright white lights low to ground behind one house. Feelings are nervous and alert. This is towards the front side of the house across the street. About an hour passes and our house appears to be under some light. Viewing out the back of the house, I notice a cold clear night with stars in the sky. A couple of stars dancing about. I notice a fog like cloud rolling over the house. The lights from town start flashing multiple colors simultaneously. The lights in town seem to be warping. My 39-year-old son says they are getting ready. I hear what sounds almost like a muffled jet engine. It appears that most of the city lights are hovering over the house in back. It appears to be a small being to front of said house. There appears to be some form of pumping in the upper window of the house which also appears in windows of a multicolored vehicle. My son video-taped the event. This vehicle stayed for several minutes. My son has another video from December 10th, 2017, which he does not remember taking. He has time loss from that time and now claims to be able to see some aliens about. These activities have been going on for the past few months. We believe military and/or civilians are assisting in some way.

Tanana, Alaska
1/13/2018 03:00
About 03:00, my friend got off the phone and said that there was a whole lot of people who had seen a space craft hovering over satellite dishes in Tanana and named a lot of the local residents that said they had been watching it together.

Fairbanks, Alaska
1/6/2018 08:25
Shape: Light
Duration:5 seconds
I was in the USAF, this could have been a dual anti-ballistic missile launch/test, as they were very fast movers, if from Eielson AFB. Looking south in the sky, from the Ralph Perdue Center, 3100 Cushman Street and 30th Avenue, Fairbanks, AK; I saw two objects appear, heading west, while climbing. They looked as if they were on the same path, but one turned north briefly, then southwest, to again trail the other. I heard a 1.5 distant roar (like a rocket thrust), as they accelerated and disappeared. There was a star, slightly above and slightly to the east of them; whereas there was also a red stationary light below this star, whereas both of these remained.

Anchorage, Alaska
12/25/2017 01:00
Shape: Circle
Duration:5 minutes
Anchorage Alaska Gray Aliens are conducting barbaric inhumane human experiments on innocent children. These needs to be destroyed! The UFOs were dropping from stationary stars over Anchorage. They were dropping straight down like they were leaving from it and the others were moving beneath the darkness horizontally. These were distinctively different than the orange circle fleets that popped up from the ground formed a serpentine geometrical danse and dispersed. Or the fake red helicopters that are sometimes utilized by them.

Palmer, Alaska
10/18/2017 09:00
Shape: Rectangle
Duration:2 minutes
Traffic was stopped on the Glenn Hwy, sitting on the bridge over Susitna River headed towards Anchorage. I looked towards the

pioneer peek and further towards the Exit Glacier. In the sky a black rectangle was standing still about level with the pioneer peak. The object would have been about 40 miles away as a bird fly. I guess it would have been huge. Then I tried to get video of it, as I am an investigator, always taking video. The camera would not focus on the object. Then it began to slope to the left and never changed from being a level black rectangle. Then it disappeared behind some clouds that were quite far away.

Nikiski, Alaska
10/10/2017 21:50
Shape: Formation
Duration:5 minutes
4 objects with rounded triangles at top. Each had separate lights in multi bulb shapes on top, each one vanished from tree line over the town.

Anchorage, Alaska
7/27/2017 05:55
Shape: Diamond
Duration:2 seconds
At approximately 5:50 am, I saw 4 bright lights over the air force base. The lights hovered in the shape of a diamond or kite for just a few seconds before disappearing. The UFO was in the clouds and did not move. Within seconds the lights fainted and disappeared.

Anchorage, Alaska
5/16/2017 03:00
Shape: Cylinder
Duration:5 minutes
Large circular object, hovering right above the tree line with white lights that blinked in a straight line, circling the object every 3 seconds or so. It was very close with no sound moving side to side in a zig zag motion and moving at a slower pace stopping periodically. Then it would speed up erratically. This occurred two mornings in a row approximately 30 minutes apart from each other on the south side of Anchorage in the same exact spot.

Chugiak, Alaska
5/1/2017 01:50
Shape: Rectangle
Duration:3 minutes
A rectangular shaped object with 4 red blinking lights at the corners of the shape followed me! I was heading home from Anchorage, and I just got back from Arizona on the Redeye. I got off the highway at the Peter's Creek Exit (east side of highway closest to the mountains). There is a parking lot with a large rock painted to look like a ladybug. I saw what looked like a low flying plane, with a flight path headed west over the parking lot from the mountains.
I pulled into the parking lot to look up and get a closer look of this "plane". As I pulled to a stop, the object slowed to a stop in front of me and vertically descended to about 100 ft. Its shape was rectangular with four red blinking lights at the corners of its shape. I remember saying out loud, "Oh my God, this isn't a plane!" After that, a white light in the middle of the object started to light up, and I started to get freaked out. I put my transmission into gear and spun around and left the parking lot quickly. It changed direction and started following me north. I started screaming because I was scared. Whatever it was, was going to try and take me. It was keeping pace with me and flying directly above me, so I increased my speed. It followed me from the parking lot by the highway exit to my street which is about almost one mile to my street. I turned north from the side road from the parking lot to head towards my street and it was still following me. I turned east down another street and then north again to head down my street. As I approached my house, I pressed the garage door opener, I slowed to a stop and started to back into my driveway. As I was backing into my garage, I lost visual of the object. For a while after that I was scared

because I was thinking "oh no! They know where I live!"

Wasilla, Alaska

3/17/2017 22:35
Shape: Unknown
Duration:2-3 minutes

Three red lights in NNW over Talkeetna Mountains, moving south for a few minutes, eventually fading. I went out to the north porch to check for the aurora and noticed one red flashing light from the north northwest, heading south. It was soon followed by a second one. I thought they were jets on a glide path but noticed there was no green wing light, or any sound, so I thought they were possibly distant helicopters. A third, much brighter or closer, red light (not flashing) came from almost directly north, and headed south, toward us, and was soundless too. It seemed less than 5 miles away. I went to get my wife and daughter to see them. When we got out to the porch, only the closer red light was visible and much dimmer even though it seemed nearer. It eventually dimmed, or perhaps, it gained altitude until it disappeared

Anchorage, Alaska

3/14/2017 23:30
Shape: Diamond
Duration:20 seconds

3 orbs in the shape of a triangle, morphed to a long thin line of light above the tree line in Anchorage. While taking the dog for a walk, at University Lake in Anchorage, we headed through a treed trail and on to a clearing. After several seconds, my daughter said "dad, look at the lights in the trees". To the north, I saw three white orbs in the shape of a triangle rise above the tree line. It appeared to turn west, and as doing so, the "orbs" turned into a long thin bright white light like a long fluorescent light bulb, but so much brighter. It then headed west, and no lights were visible. I opened my eyes wide to let as much light in as

possible (the moon was not above the horizon at the time). That is when I noticed the shape, very much like a box kite, but with triangles on top and bottom, essentially an elongated diamond. We stood there for about a minute after and headed back down the wooded trail talking about what we had seen. After about 40 yards, I said let's go back. I was shining my flashlight on the tree line and my daughter said that if we see anything, she will record it with her
I-Phone. Once again, three orbs in the shape of a triangle appeared in the same spot slowly rising above the tree line. This time my daughter took out her phone to record and her phone died. She said it was at 60%. The same scenario as above played out, the orbs turned to one long bright light, then they headed west. It was just a dark silhouette of an elongated diamond until it went out of view (total time about 20 seconds each occurrence). At this point we left as it was cold. My daughter's phone did not turn back on until we got home and plugged it in. Immediately it showed 62% of battery power.

Anchorage, Alaska

3/14/2017 23:30
Shape: Other
Duration:20 seconds

Trapezoid shaped object rising above trees leaving to the West over Anchorage.
This experience was played out as follows. My dad and I walked to a large field where we saw the unidentified craft, we saw it flying through the tree line like the trees were non-existent. It was the second time the craft came around that I pulled out my phone to start recording it, but my phone had died, and it was at 60% (my phone could never die with that much power left.) The craft seemed to patrol in the same manner as it circled around the snow-covered field. The shape was very odd, having a trapezoid-like shape with lights at the end of every point. Another detail I noticed was the

peculiar hieroglyphic design on the bottom of the craft. The design was on plate with the mysterious hieroglyph etched into it.

Soldotna, Alaska
3/4/2017 04:00
Shape: Unknown
Duration:1-2 minutes
Bone-rattling sonic boom like noise that did not "fly by," but ascended directly upward, no visible craft seen. I was on the peak of a one-story house roofing, when I heard what seemed like a sonic boom something. Living in Alaska, I have only experienced this maybe 2 times. I remember distinctly experiencing the first as a child and how it led me to ask and learn about the speed of sound, etc. After hearing a massive jet fly over yet looking up to see the only aircraft in the sky was already far past me, I, so on this super chilly yet beautiful bluebird, (literally not a single cloud visible in the sky) sunny day, I was sitting on the peak of the roof and as with a sonic boom, this out of nowhere, thunder-like booming began. I started to scour the sky for the plane. There is a very small municipal airport nearby, so watching the various and at times quite suspicious types of helicopters and planes that come and go, is something I actively do as we work. As the intense sounds continued it was immediately obvious that the sound was not "flying by" but instead was isolated almost directly above me, a little more towards the northeast though. It was if a rocket was taking off above my head but that it was either so far up that my (extremely excellent) eyes were unable to see it or that whatever it was, was completely invisible. I have read the unaided human eye can see a candle flame from 30 miles, so whatever was producing the noise was not reflecting light, was behind a barrier or was projecting its sound "angularly" by bouncing it off of a barrier. It was so loud that it literally rattled my bones with cracking and booming, freaking me out. Another way to

describe how the whole thing felt is like the blue of the sky was the surface of water, and I couldn't see up into it, but right on the other side of the surface was an enormous turbine propeller that literally tore its vessel away with power resulting in what sounded like a sheet of tin, as big as the sky around me, being shaken over and over and over like you do with sheet to make a bed. I searched the sky for any movement for minutes even after the sound ended but saw nothing. I have had two unidentified sightings before this. The first in late Dec 2013 shared with another witness and the second not reported from winter of 2015.

Fairbanks, Alaska
3/3/2017 21:38
Shape: Sphere
Duration:1-2 minutes
Large glowing sphere makes course change at high altitude.
As I drove through an apartment complex on Davis Road, I noticed an unusual bright glow approaching from the south. I immediately stopped my truck and started recording video. I pointed it out to my passenger who was also stunned. The object was moving from south to north, before changing to an eastern course and disappearing behind a cloud layer 7000-10000 feet. It was difficult to judge distance, but I believe the craft was at a much higher altitude. Likely massive size.

Anchorage, Alaska
3/3/2017 08:15
Shape: Fireball Duration:10 minutes
The object was observed to the east of the city of Anchorage.
The object was reported observed over or beyond the Mountains to the East of the city. Cool morning, approximately 11° F at ground level, clear morning sky, light breeze.
The object appeared vertical at first, then horizontal and proceeded to travel to the south at a rather fast pace. No reports of

sounds or booms, and unable to determine distance, other than sight of the object was lost as the mountains blocked the view briefly behind the peeks.

The object was bright as it was back lit by the sun, and the western face of the mountains was still shadowed. There was a trailing plume that was visible and changed slightly in form the trailing edge did not linger and closer to the tip remained constant; however, the leading edge appeared sharp and constant.

Wasilla, Alaska
2/26/2017 22:51
Shape: Oval
Duration:2-3 minutes
3 low and slow-moving pulsating orange/reddish orbs.
3 round objects that were a glowing orange color that pulsated. No position lights as on civilian or military aircraft. Very low, thick cloud layer at approximately 1000 feet above ground level. Objects moved slowly and methodically. Was about 1/2 mile away, approaching my position from the southwest. 1 broke from the loose formation and headed south and then rose into the low cloud layer. The other 2 were in a staggered right formation. Lead object stopped and seemed to wait for the other to catch up and when it did, they both went straight up in the cloud layer.

Fairbanks, Alaska
2/17/2017 20:07
Shape: Circle
Duration: 8 minutes
Several bright orange objects, moving east, semi sporadically, most lined up at one point, disappearing from sight one by one. Have pictures and video

Wasilla, Alaska
1/20/2017 22:00
Shape: Light
Duration:5-10 minutes
Bright light in the sky like a lighthouse beacon. I was driving home from work one night, and as I was driving down the Parks Highway in Wasilla, Alaska. Up in the sky, I noticed a bright white light. While observing this I estimate that it was roughly 3-4 thousand feet up just above the cloud layer. This light was like a lighthouse beacon that revolved about once every 5-10 seconds. It stopped after about 5-10 minutes of me noticing it.

Anchorage, Alaska
1/1/2017 00:10
Shape: Formation
Duration:3 minutes
Red round lights in 5 or 6 groups of 3 lights each hovered over Hillside area during midnight fireworks seen and from Tikahtnu area. From the viewpoint of the north end of Muldoon area near the Glenn Hwy, at 10 minutes after midnight, we saw 5 or 6 clusters of what looked like small orange steady lights, 3 to a group, each group of three kept a perfectly tight triangle pattern formation. They came up from the south at a medium speed, too slow for a firework rockets or flares, but faster than most airplanes. They slowed to a stop still in formation still due south. It was hard to guess the distance being dark, but my guess would be about 500-800'in altitude and maybe 5-8 miles due south from us. The groups of three seemed to keep perfectly together within each group as if there was one object with three lights each but the different groups of three would vary slightly in position relative to the other groups. Once they stopped, they were still moving slightly around but staying in the same rough area. There were no flashing lights, but the orange balls seemed to slightly pulse with a subtle color shift from almost yellow to deep reddish

orange. After watching for 2-3 minutes, I ran inside to grab my camera but when I got back less than a minute later, they were gone or moved too low to see due to a 2-story apartment blocking my view. At first, I thought they were part of the neighborhood fireworks, but after watching, it was clear that they weren't. We were joking about the fact that they seemed to act like they came in close to watch the midnight fireworks show, then left when the fireworks were thinning to a close. Also, they made no sound that could be heard. That could have been because of the distance fireworks' noise.

Anchorage, Alaska
12/31/2016 23:50
Shape: Fireball
Duration:20 minutes
Three bright orange orbs, exhibiting odd flight patterns, above Ted Stevens Airport.
Three hours after witnessing one large, bright, orange orb travelling south-westerly over mid-town Anchorage, I sighted upon three identical orbs a few miles (in my estimation) above the Ted Stevens International Airport property. These were moving around in separate, oblong paths at casual speeds, with sometimes one or another momentarily breaking away from its path to perform a zigzag or to move somewhat farther afield, and every few minutes all of them slowly dimming and returning to full brightness. After about 20 minutes of watching, they simultaneously dimmed out for good as a routine aircraft was making its approach from the Chugach Mountains.

Delta Junction, Alaska
12/23/2016 18:10
Shape: Fireball
Duration:45 seconds
Seven fireball glowing/pulsating objects were seen moving in erratic, non-linear fashions over Jack Warren Rd at night. They were in a shape similar to the Big Dipper. They made no

sound. They were moving slowly, kind of bobbing high in the sky. At the end of the sighting, each object left a streak, kind of like entering a warp/light speed. I am a navy vet who worked on aircraft. These objects did not move in a fashion consistent with any hovering or moving, jet, helicopter, or missile. The objects emitted no lights in the fashion of aircraft (strobe lights/navigation lights)

Chapter 6

The Great Lakes Region

The Great Lakes are the largest body of fresh water on Earth, being one fifth of the Earth's fresh water. Lakes Superior, Huron, Michigan, Ontario, and Erie are a combined 6 quadrillion gallons of water. These lakes Border, Michigan, Wisconsin, Minnesota, Illinois, Indiana, Ohio, Pennsylvania, New York and Ontario, Canada. Lake Michigan, sight of the Michigan Triangle covers over 300 miles of water and is one of the most mysterious bodies of water on Earth. Numerous reports of unidentified craft and balls of light hovering over or coming out of the water, are reported to MUFON, International (Mutual UFO Network) monthly and investigated by trained MUFON Michigan Field Investigators. Reports of unidentified craft are also reported monthly to NUFORC (National UFO Reporting Center). Some of these witness reports are listed in the next few pages.

MUFON Witness Sighting Accounts

Chicago, Illinois
Case#64767
October 15,1953 10:00PM
Witness and 2 friends were coming from a church function walking home, became aware of something over their heads. Looked up, saw a white disc which was bright like a new moon. There was a strange indentation on the bottom of the disc, it was silent and gilding over their heads and buildings headed toward

Montrose Harbor at Lake Michigan. It disappeared as it descended toward the harbor. A week later a boy in class described seeing the same object the same day. The other boy was 2-3 miles away

Lake Michigan Chicago, Illinois
Case #67184
September 7,1962
Teenage couple parked at Montrose Harbor spotted a flying saucer coming towards their car. It traveled from Lake Michigan going West. The craft stopped and hovered directly over their car. The witnesses looked up saw a white domed flying saucer. The craft had white lights underneath it turning to red and green lights with identification marks on the underside. It moved in a silent gliding motion over 500 foot above their car. They got scared and took off immediately. Numerous people called the newspapers to reported flying saucers. Her mother and father also saw the craft headed west over the city.

Oakville, Ontario
Case #81347
March 1, 2009
Was driving through Oakville, ON at 5:45 am traveling down Reynolds Street headed toward the lake. The witness noticed what looked like a plane come towards her, lights flashing. The lights suddenly stopped flashing. They appeared to be too low to the ground for a plane. No sound. It stopped, hovered over a 10-story apartment building. She drove under it and could see 4 lights. It appeared as though

driving under a deformed rectangle. The craft was massive. She drove away as fast as safely possible. Nobody else was around.
Experience: 5 mins or less

LAKE HURON
We were camping on September 12, 2009, at Lake Huron Campground, at 9:25 PM, and noticed four very bright red/pink lights. The four lights were in a perfect geometric form similar to a cross and did not blink or lose their form from the time we saw them until they moved from north to south. They flew toward us in perfect formation then suddenly the two center lights dimmed and disappeared. The top and bottom lights went out at the same time. We thought that the lights may have been flares, but a planes altitude would have been very low to drop them, and we heard nothing. Also, the lights did not flicker or vary from their exact formation. If they were flares, the lake breeze would have caused some shift in formation and flickering.

After the red lights went out, we noticed a dark shadowy object moving east faster than any jet from where the red lights had been. As it moved east, it got lower and flew over Lake Huron. It was three treetops in size as it descended towards the water, and we saw flickers of white light directly underneath the object similar to a reflection of light that was shimmering off moving water. It kind of sparkled under the object. The object was darker than the starlit night sky and appeared as if covered by black camouflage fabric. It made absolutely no noise and appeared to vanish either onto or into Lake Huron. My husband and I both have master's Degrees; we're intelligent, reputable individuals.

Lake Michigan, Chicago, Illinois
Case #60001
Sept. 14, 2014, 1:00AM
The witness and friends were staying at a 15th floor apt that overlooks Lake Shore Dr. and Lake Michigan. They were on the balcony and noticed a bright orange orb ascending from the lake. This was first seen approximately one mile out onto lake, and a couple hundred feet in the air. The object was rising diagonally upwards at a slow yet keeping a steady velocity. It stopped and hovered, then began moving east. Additional orange orbs were seen following same flight path. It seemed as if there was a fire beneath the surface of lake from which the orbs rose. This occurred four more times over the next five minutes. One orb seemed significantly closer and faster than the rest. There were no lights from boats on the lake. Vantage point was 135 feet

Toronto
Case #64162
March 24, 2015, 9:15PM
Clear evening, went out for a smoke, observed a dimmed V shaped object over head at 3,000 ft. The outline was of a V shape. There were 7-9 round yellow lights not blinking.
Estimated speed: 150 mph
No engine noises.
The craft was moving south on Dufferin Road
Sighting 45 sec.-1min
Craft flew out of view from south to north
MUFON Investigators: Dave Palachik & Patricia Hart

Stoney Creek, Hamilton, Ontario
Case #69754
Aug. 22, 2015, 8:00pm
The witness was walking the dog, noticed two strange lights glowing. The Lights began moving towards the witness, stopped, and hovered. The witness began walking towards the lights which moved away in the opposite direction. The lights then slowly began to fade until eventually they disappeared. Minutes later both lights reappeared at a standstill hovering in the sky. A group of teenagers at a school park began to watch the lights as well. The witness decided to quickly walk home, drop off the dog, and then get in a vehicle

heading up a hill to get broader view of the city. The lights appeared everywhere, there were 20-30 orbs of lights at eye level. The lights were northwest over Stoney Creek and Hamilton skyline. They resembled frozen fireflies. The Lights began to move making various shapes and patterns in the sky. 20-30 lights then decreased to around 10. At one point only three lights were visible forming into a triangle. The lights all reappear and formed into what looked like DNA structure. The lights were continuously disappearing and reappearing reforming shapes. The witness tried phoning family and friends but no signal. Sighting lasted approximately 30 min.
MUFON Investigator: Marie Malzahn

Stoney Creek, Hamilton Ontario
Second Sighting
Case #69699
Aug. 22, 2015, 8:15pm
Witness was walking home from work, same route as always. Turned the corner and could see the sky in plain view. There were 10 to 15 glowing lights similar to stars but moving. A dark storm cloud was in the background. The cloud helped to contrast the glowing lights. The lights hovered in way that appeared similar to someone investigating a storm cloud. The lights floated around, at one point they almost began to join into a formation, before slowly dispersing one by one into the storm cloud.

Benton Harbor, Michigan
Case #91316
April 9, 2018
The object was also seen several other days in the northwest sky. It was brighter than Venus. The sighting started about the time Venus was setting. The object was cigar-shaped and made no noise. At one point, it did change position slightly. It was not celestial, did not track with the motion of the stars or set in a similar way.

Sighting lasted at least 2 hours
Investigator James Wolford

Historic Sighting Reports
"UFO-USO and Extraterrestrials of the Sea"
Maximillian de Lafayette
Carl Feindt for providing data
"Invisible Residents" by Ivan Sanderson

08-21-1867 Lake Ontario, Canada: Hamilton Ontario Times, 08-23-1867
08-LL-1914 Near Georgia Bay, Canada: Two accounts- John Magor/ Timothy Beckley
04-25-1946 Anima-Nipissing Lake, Ontario, Canada: The Northern News
10-14-1948 Bear Creek Harbor Lake Ontario, 1:15am fiery ball, burning plane going a terrific speed moving up and down seen by 2 adults
09-16-1957 Lake Huron near Port Huron, Michigan, USA: Hammond Times 09-16-1957
04-08-1958 Elyria, Ohio, USA Lake ERIE: Confidential NICAP Bulletin 07-09-1958 7:20PM fiery object with 2 horns, sparks like molten metal from front and sides seen by hundreds
05-24-1958 Lake Huron, Grand Bend, Ontario, Canada: Invisible Residents by Ivan Sanderson p. 226 1am red flare arched into sky leaving trail of red light then plunged, Prov. Police
10-13-1958 2am Wayne Co. NY, Lake Ontario, fiery object, burning plane, crashed
10-14-1961 Lake Superior off Two Harbors, Minnesota, USA: The APRO Bulletin Nov. 1961 page 3. Object the size of an ore carrier, skidded into water, bobbed on surface, then took off SE
12-02-1962 Lake Erie, Akron, Ohio to Syracuse, NY, USA: Invisible Residents by Ivan Sanderson p. 226 Bright lightbulb shape object traveled east angling to north, arched into lake
04-21-1963 Lake Michigan off Chicago, Il., USA: Chicago Tribune 04-22-1963
12-02-1963 Lake Erie off Dunkirk, NY, USA: The APRO Bulletin Sept. 1964 pg. 6

Glowing object passed through sky, plunged into lake, coast guard search

02-14-1964 Lake Michigan in Michigan/Indiana: Invisible Residents by Ivan Sanderson page

10-10-1964 Lake Erie Port Colborne, Ontario, Canada: Invisible Residents by Ivan Sanderson page 226. Flaming object fell from NE faster than flare, slower than meteor Crash heard, nothing seen, several witnesses

11-07-1964 Lake Michigan, Michigan, USA: Invisible Residents by Ivan Sanderson p. 228 1:38am flash of light bolide

01-21-1965 Lake Michigan, Michigan & Illinois, USA: The Chicago Tribune 01-22-1965 5:40pm possible meteors, appeared to explode and fall, seen by many including police

08-04-1965 Louis Bay, Duluth, MN, USA: Invisible Residents by Ivan Sanderson p. 43

09-05-1965 Lake Superior, off Marquette, Michigan, USA: Invisible Residents by Ivan Sanderson page 228 Fireball plunged into lake 10 miles offshore, coast guard & others

11-27-1965 Lake Michigan, Racine, Wisconsin, USA: Invisible Residents by Ivan Sanderson page 228 Brilliant small light grew to large size, appeared to separate before it disappeared, many

12-09-1965 Ohio/PA PM hour, fireball dropped into lake, seen by thousands

12-12-1965 Lake Erie, OH & PA, USA: Canton Repository, Canton, Ohio, USA...same as above

03-14-1966 Lake Michigan, Chicago, IL. USA: Invisible Residents by I van Sanderson p.228 9pm fireball, moved from east to west and disappeared,

06-01-1966 Lake Ontario, Canada: UFO Investigator Vol. 3 #9 Aug-Sept. 1966 p.5

09-11-1967 Vermilion, OH Lake Erie, yellow object possible meteor, visible about 12 seconds, may have split in two or may have been two objects, seen by several

11-XX-1967 Lake Ontario, Kingston, Canada: Grass Roots UFOS by Michael Swords p.37

04-17-1968 Georgian Bay, Ontario, Canada: F.S. Invasion by Steiger & Whritenour p.96

11-24-1968 Lake Erie off Vermilion, OH, USA: Invisible Residents by Ivan Sanderson p. 228 PM Hours, plane crash, fireball, craft skidded along water until flames disappeared. No sound, smell of diesel oil, seen by three persons including two boys

06-20-1969 Lake Michigan offshore Chicago, IL., USA: The UFO Investigator Vol. 5 #1 Sept/Oct 1969

07-06-1973 Lake Erie off Cranberry, PA, USA: The UFO Register #366

09-06-1973 Lake Michigan off Sheboygan, WI. USA: Newspaper Clipping, Sheboygan Press 09-07-1973

08-XX-1981 Hamilton Harbor, Ontario, Canada: Investigator MUFON ASD, DE. Carl Feindt

04-XX-1993 Lake Huron off Kincardine, Ontario, Canada: Brian Vike, Dir. HBCC UFO Research 09-03-2006 White Fish Bay, Michigan, USA: MUFON CMS

MUFON Sighting Reports United States and Canada

Chicago, Illinois

Case #12288

08-16-2008

I was watching the 2008 Chicago Air and Water Show at Lake Michigan. It was 16 August, and I was at Fullerton Street about 30 feet from the water. I was watching a Navy F/A 18 doing maneuvers. I was using binoculars to watch planes when they went higher or further away. The F/A 18 did a maneuver that took it vertically straight up in the sky and then it did a stall holding its position for a short while. As it was going up, I used my binoculars to keep track of it. While I was watching it, I noticed two objects in the sky above the plane. They were hovering motionless, and I was surprised to see them. The distance from these objects and the F/A 18 was hard to tell but they were not very close. But the pilot of the F/A 18 would have seen them since he was headed right for them but did the stall before getting too close. I would be very surprised if he did not see them. Upon first noticing them I tried to figure out what they were. My binoculars were not strong enough to give me the greatest detail, but I could make out what they looked like. They were probably a few hundred yards apart. The closet thing I can compare them to are the rovers that landed on Mars. They resembled the bubble-like structure that shielded the rovers as they landed. There were bubbles, as I will call them, all over the object except the bottom. The bottom was triangular and black. the points on the bottom were rounded. The bubbles were cream colored. There was no visible sign of a propulsion system that I could see. They stayed in that position for about five minutes maybe a little longer. I would watch them and then look at the F/A 18 as it continued its maneuvers. I held my binoculars in place as so I wouldn't lose track of them. Each time I looked back they were there and hadn't moved. I looked around the crowd to see if anyone else had noticed them, but I didn't see anyone who did. As the F/A 18 was ending it's show I watched as it flew away. I still kept my binoculars in the same position but when I looked back this time they were gone. I scanned the sky but could not locate them. I still don't know if anyone else saw them, but there was a video crew there taping, and I would not be surprised if they got something on tape. I myself had no camera and I wish I did.

Case #30170
06-28-2011
I was at Oak Street Beach in Chicago on Tuesday, June 28, 2011, between 11 am and 1 pm. While relaxing and lying down on the beach near the water, I saw a white

orb/circle/disc glide out of a white cloud in the east over Lake Michigan. It was in the area of the sun because I had to block the direct sunlight in order to see it. I watched as it traveled quickly at a downward angle to the south until it was out of view. This happened at least three more times in the same area. When I tried to record with my phone, I didn't spot any more orbs. I watched eye floaters, birds, helicopters, and airplanes that same morning, these were nothing like any of them. They looked just like the London orbs recorded recently.

Case #60001
09-14-2014
Some friends and I were staying at a friend's apartment for the weekend which overlooks Lake Shore Drive and therefore Lake Michigan. We were on the balcony having a conversation when I happened to notice a bright orange orb ascending from the lake. I quickly alerted the two other people who were with me to the object and none of us could identify it. When we first noticed it, it was perhaps a mile out onto the lake and a couple hundred feet in the air. It was rising diagonally upwards at a slow yet steady velocity. Once it reached a certain altitude, it appeared to stop and head East, at which point the brightness of the object quickly diminished. Once the object disappeared, we resumed our conversation, only to be interrupted by an identical object rising from the lake which followed the same flight path as the last object. Once this object disappeared, we continued to watch the lake to see if another object would appear and if we could pinpoint its origin. Sure enough, about a minute later it looked as if someone lit a fire beneath the surface of the lake or on the surface of the lake from which another orange orb rose, once again following the same flight path until it faded away in the distance. This happened 4 more times over the course of the

next 5 minutes, with one orb in particular seeming significantly closer and faster than the rest. My initial thought was that someone was on the lake lighting Chinese lanterns, but I was forced to dismiss this idea for several reasons. Each object followed the EXACT same flight path and seemed much too large be a Chinese lantern. There were also no lights from a boat on the lake where somebody could be lighting them off, which from our vantage point we would have most likely been able to see. Finally, who would be on the lake after midnight lighting Chinese lanterns? After the event we all went inside and prepared to go to sleep, all of us still wondering what we had just witnessed.

Case #64767
10-15-1953
I and 2 friends were coming from a church function. As we were walking home at approx. 9pm to 10pm October 1953 or 1954. We became aware of something over our heads. As we looked up there was a white disc overhead. It was bright like a new moon. There was also a different indentation on the bottom of the disc.   It was silent and gilding over our heads and the various buildings headed toward Montrose Harbor at Lake Michigan on Chicago's north side. It disappeared as it descended toward the harbor. We called the aeronautical civil service board. We then could hear a light plane fly in the area. We assumed they sent up a spotter plane. We knew what we saw!! About a week later during a science class in school, we were reporting what we saw when to our surprise, another boy who was not with us and at least 2 to 3 miles away also said he saw the same object. He descriptive it in the same way as I have related. We all knew that this was a saucer we were seeing. The next summer in the same area of the Lake Michigan Montrose Harbor, there was an event in the lake and at Montrose Harbor entrance. A tidal wave 100 ft

tall which do not happen in lakes, swept at least 30 people off the jetty where people were fishing. There was witness to that event and for weeks they were finding bodies. I always thought that these 2 events were tied together somehow. After watching a program on UFOs in lakes it seemed to match.

Case #67184
09-07-1962
We were sitting in my boyfriend's car talking. We looked up through the front window and above the car was a hovering spaceship like you see in movies. It was very large with lights blinking in white then red and green. A minute after we saw it we were just about to leave because we were afraid. Then it just went from where we were to way out over the city in a second. Even though the government had radar installation sights on the lake. No sound no fumes. Just from point A to point B. There is no way it could have been anything that our government had made. When we arrived at my boyfriend's parent's house, we were going to tell them about what we saw and his mother saw the same thing in the sky above her home on Milwaukee Avenue, which is directly west of Montrose Harbor. Others reported on the radio that night seeing a UFO that night.

Michigan

Center of Lake Michigan
Case #22658
05-22-2005
A group of 3 of us, myself, a manufacturing business owner, my brother a Firefighter/Paramedic, and a family friend were traveling from Saugatuck, MI on a 38' Chris Craft yacht, crossing Lake Michigan on the way to our marina just south of Chicago in Burnham, IL. This was a 15-hour ride, and we were about 6 hours into the trip when we were circled by an F4 jet several times at very

low altitude. This jet was so close we could clearly see the pilot in the cockpit (about 40 feet over the water). I do not know if this is related to what happened 30 minutes later, but we all found it very odd to see an Air Force jet in that area taking a closer look at us. Of all times I've been on the lake I've never seen a fighter jet or anything odd, and I've been out there since 1994. Eventually the jet vanished and was never to be seen again. About 30 minutes later we noticed an odd- looking object to our starboard side (we were traveling SSW at the time). This object was nothing any of us had ever seen and didn't look like the commonly known UFO either. It was not a cigar or disc, and it was not a triangle. It was the strangest looking thing. It was completely black and almost looked like a bunch of LEGOs a child threw together, where the long rectangular shape had odd ends like steps but completely uneven. It's very hard to describe. There were no lights, no sounds, just complete silence. We idled down and proceeded to the bow to observe this thing. It was hovering above the water, about 10 feet above the 1' wave caps. It was about 500 feet wide, almost like a large building on its side floating over the water. We were about 6 blocks to 1 mile away. After about 10 minutes of watching this thing, it began to move our way, and we were all getting a bit nervous being that we were in deep waters and all alone. Outside of cellular range and poor chances of a good VHF distress call, our feelings were becoming uneasy quickly. I guess we just feared this object because we didn't know what it was or why it was even there. I hopped back inside the cabin to take the helm and threw her port and starboard engines in forward at idle to get ready to start moving out of there. I wanted to know what I was seeing but I was way too nervous to see what would happen if I let it come close. At that moment I myself, along with our family friend and my brother, observed this crazy looking thing move across

the water from our starboard side to directly in front of our bow (keeping the same 6 blocks or so away from us and holding the same 10' hover over the water). What really shocked us was the speed of the object. It traveled an estimated 10 nautical miles in about 2 seconds. There was just a long black streak across the horizon and then it stopped and just sat there in front of us. I dropped the vessel back to neutral and just sat there as we all discussed what the heck was happening. Seconds later, it jumped right back to our starboard side where it was before and again just hovered over the water. That was enough, we throttled up and began heading SSW again, constantly observing the object as we moved on. Eventually we lost the object in the distance and never saw it again. The date was, for sure, May 22nd, 2005. That's the only time we ever made this trip. We had no cameras, have no photos, but it happened. The trip was being made for a single purpose, to move a boat I bought at Tower Marine in Douglas, MI to my local marina. It wasn't a pleasure cruise. In June of this year, I am returning to the GPS coordinates where this happened, still saved on my GPS. I'll be going back with a couple friends and possibly my brother again if he can make it. My hope is to see this again, although it's been 5 years now, you never know. This time we're bringing camcorders and cameras.

Scottville
Case #27831
02-14-2011
I was looking in the north sky over Mason County and noticed a triangle formation that I thought was three airplanes, the front one was a solid white lite with a blinking red, and the back two were blinking a reddish orange light. They were heading west and once they got above Lake Michigan the front one curved south, and the two in the back disappeared. When the front object was going to the south

it was just a star like appearance without the red blink.

Bay City
Case #38269
08-15-2011
It was approximately the middle of August 2011. My family and I were on our boat going down the Saginaw River heading towards the bay. It was extremely foggy out and also there were thick clouds in the sky. We reached the mouth of the bay and the fog started to clear out in the water but was still heavy on the shore. I was looking at the fog line when all of a sudden, I saw something move in the fog and clouds. It moved in and out of the fog and clouds and appeared to be hovering but moving slightly back and forth. It was a dull metallic dark grey rectangle object. I could see what I thought to be red lights on it. We were approximately 500ft from the shore, but I could see it plain as day peeking out of the fog and cloud cover and then going back in. I watched it for as long as I could see it. I told my husband to look, and he was busy driving the boat and either didn't hear me (our boat is loud) or just didn't look. I told my kids to look but they were young (then 5 and 6), and they didn't see it. This was the first time I saw this object, the second time, which I will also post, my daughter, then 7 years old did see it too. I didn't know what it was then and still don't know what it is now.

Coloma
Case #43787
10-17-2012
My girlfriend and I were taking a walk on the beach at night watching the stars. While sitting on the beach, we both noticed two circular objects above the water. At first, we thought they were stars, but suddenly they started moving above the water in different directions. That is when we started paying close attention to these star- like objects. At night it is normal to see planes flying high in the sky leaving or arriving in Chicago. All these planes fly very high, but these objects were flying very low above the water and were moving at very high speeds. We didn't know what to think about these objects, but things really got strange when all of a sudden, they were six of them. After serving several years in the military and witnessing all types of planes and helicopters in flight, it was apparent that we were witnessing some very special type of flying machines. We were both excited and could not believe what we were witnessing, but it was happening, and it continued to happen for about half hour. They all seem to surface from the water. We never notice them come from the sky down, but up from the water. We lost sight of them after they seem to go down into the water. In the eight years I have walked on that same beach at night, this is the first time I have ever witnessed anything like that.

Ludington
Case #90849
07-29-1989
I deliver yachts on Lake Michigan. At dusk in the geographic center of Lake Michigan, the last weekend in July of 1989, I was with four other guys. We saw 5-6 bright aqua-colored discs under the water, moving in a zig-zag pattern. After 3-4 minutes, they came out of the water, hovering over the surface. All of a sudden, they shot straight up into the sky, and disappeared in a blink of the eye. Today, two of the guys admit what we saw, the other two will not. I have been on all the Great Lakes delivering boats for years, and Lake Michigan is the only lake on which I see weird stuff. In 1996, closer to Chicago, I saw white lights under the water that zig-zagged and hovered too.

Benton Harbor

Case #91316

04-09-2018

This has been seen for days if not weeks now. Looks like a very bright star hanging low in the northwestern sky. Brighter than Venus, which will set. This does not set. There is an airport out that way and sometimes planes with bright headlights do look like UFOs, so I haven't paid much attention to this until about a week ago when I just watched it and unlike the planes this never moves. Planes all eventually overhead, and they always make noise. This thing just hoovers there for hours. A few nights ago, I lined up my sight against a fixed object and this thing actually tracked east ever so slowly. It is like watching the movement of the moon against say a pole. Tonight, I dragged out my old telescope. In the scope this thing is not round at all. It looks like a dash in the sky, almost cigar shaped. I got a few blurry shots because my scope has fallen down a few times so it might have issues. I tested it earlier by focusing on houses down the road in daylight and was able to focus it just fine. The sky was still too light to see stars. I got it as sharp as I could, then just held my cell phone over the eyepiece. One photo shows one of our power lines in the field almost covering part of this. Saturday my neighbor, my daughter and I were all outside watching this thing. The night before it was further north and lower in the sky. Last night was too cloudy and tonight it's back to its usual west-northwest spot. I have not witnessed where it goes but at some point, it just vanished My neighbor told me it just disappeared one time he was looking.

Wisconsin

Milwaukee

Case #10467

11-30-2007

Strange object over Milwaukee Harbor, two pictures submitted

Asland

Case #10744

05-XX-2000

My wife and I come from New Brighton, MN, which is a northern suburb of the Twin Cities (Minneapolis/St. Paul). Every few months for the past decade, we make short vacation trips up to north-central Wisconsin, or to the north shore of Lake Superior. We enjoy the beauty of the lake and country around the area of Ashland and Washburn, in particular. When we travel to north-central Wisconsin, we almost always stay overnight on the east side of town, next to Lake Superior. On a May morning about eight years ago, I got up and went out to the north patio of the hotel to sip some coffee and smoke a cigarette. As I lit my cigarette, I scanned the lake for boats and ships. My attention was immediately attracted to what I at first took to be a bird spiraling up from the surface of the lake. I hadn't seen the "bird" start from on the water, but I noticed it at a low altitude and rising. The "bird" was rising slowly, flying in a tight upward spiral, counterclockwise as seen from above. After a while, I became increasingly puzzled, since the "gull" wasn't deviating from the tight upward spiral, as one might expect from a bird. Also, I noticed that the object was most brilliant when it passed through the west side of its upward spiral. Curious, I watched the object for some minutes, as it continuously spiraled upward. Eventually I lost sight of the object at an elevation of some 70 degrees, owing to eye strain and growing distance (altitude). Using a simple tangent formula, and assuming a base distance of about a mile, the "bird" would have reached an altitude of tens of thousands of feet. This seems strange behavior for any bird.

Kewaunee

Case #27755

02-08-2011

We were coming out of work which is located on the north side of the city of Kewaunee. As we walked across the parking lot someone yelled to me to look up in the western sky and tell them what they were seeing. Looking up, I saw a gold, spikey, shaped ball coming towards us at approximately 1000 feet. Another worker came up and she and I saw the object turn towards the southeast and accelerate very quickly. Reaching our location and now being ball shaped, looking at it from the side, the goldish blue light suddenly disappeared, and two tiny white lights appeared to flash on and off taking only a couple of seconds to do so. After that it was gone from our sight. The weather at the time was clear with numerous stars being seen. Temperature about 10 degrees. Object was seen by 3 people. Duration approximately 30 seconds. Feelings were more of curiosity than anything else

Green Bay
Case #78896
09-03-2016
While sitting in my hot tub I always watch for satellites. I noticed 1 bright white object that was darting around in multiple directions then would stop and hover. I then saw 3 more in different parts of the sky. The first object was over the northern bay of Green Bay, one to the south of me, and two to the north. I grabbed my binoculars and could see bright white orbs with multicolored pulsating lights. They would move abruptly in any direction they wanted to, then stop and hover again. These objects remained in the sky for several hours. They had absolutely no sound. I estimate their elevation at 15 to 20 thousand feet.

Little Suamico
Case #80998
11-28-16
We saw a triangle come out of the Green Bay. It made a muffled jet sound and was traveling at a speed of 35 knots, too slow for normal lift versus air speed as we know it. It had one light

on each corner. The size was 100-150 feet in length.

Canada, Ontario

Pickering
Case #44582
10-22-1992
On October 22, 1992, about 11-11:30 pm, a friend and I were sitting in a car on Pickering (Ontario) beach, facing south looking out over Lake Ontario, just west and adjacent to the Pickering nuclear power plant. This object came out from behind the plant over the lake, flying in a straight line, from east to west. It was a cylinder-shape, with a rounded nose that glowed red, the body of the object appeared a sparkly emerald green, and something that looked like a shimmering blue (the color of a gas flame) came from the rear. At first, I thought it may have been a meteor, but seemed to be moving too slowly. The object flew straight, it did not get smaller, it did not dive down into the lake or disintegrate or break up. It simply vanished right before our eyes. I was dumbfounded, but my friend and I both saw the same thing. I have often wished that I had contacted the nuclear power station immediately afterward, as I am sure they must have caught it on security cameras, but I was younger then, and afraid of ridicule or being called a "crackpot"!

Goderich
Case #60268
10-01-2014
One object with halogen bright red, green and white lights hovered N, NE over Lake Huron approximately 45 degrees, another was spotted at approximately 90 degrees same description moving back and forth in the sky. Both objects were spotted at the same time. I have seen this object in the sky on occasion since June. At times the sighting will last from

2-3 hours. I noticed it tonight at 9:10 and it was gone by 10:00.

Owen Sound
Case #62844
07-17-17
My friend and I had camped on the eastern side of the Bruce Peninsula on the western shore of Georgian Bay a few miles north of Cape Chin. In the evening at around 7 p.m. we looked to the southeast and observed an extremely bright light moving over the water at a moderate altitude of perhaps a few thousand feet. The light moved at a very high speed and then immediately stopped. It changed direction and accelerated at a tremendous rate stopping abruptly again. It repeated this pattern a number of times. In one of its maneuvers, it appeared to be coming toward us. I admit that I was a little frightened as I had heard of alien abductions. We watched the object for perhaps ten minutes. I don't recall exactly how we lost sight of it. There were many reported sightings of this object that evening. Most were on the southeastern side of Georgian Bay where there are thousands of summer cottages as well as a number of towns. My friend, a lawyer, and I felt this was not an aircraft of any kind as it moved so quickly and erratically. I possess a commercial helicopter license and own a Hughes 300 helicopter and I have a Bell Jet Ranger endorsement. I know how helicopters perform and there is no way a helicopter could do the maneuvers we observed with such tremendous acceleration and abruptness in stopping.

Toronto
Case #64162
03-24-2015
This is my third unexplained sighting since September 2010, the last two were in the last two evenings, March 23-24, 2015. This is directly from my notes I made immediately after my sighting. I typed these into my phones note app, emailed it to myself, and copied them directly here. V shaped UFO sighted at approximately 9:15pm, on Tuesday, March 24th. A perfectly clear evening, I came out for smoke. I always look up, caught the dimmed V shaped object as it flew overhead at approximately 3000 feet give or take a thousand. I could see the distinct outline of a v shape with approximately 7-9 round yellow lights that were not bright at all, as if the craft was encased in a dark mesh like cloak. The lights appeared to be like 10-15% brightness. I say it was moving at a clip of 200kms to 300kms. hard to tell, but like a little faster than an aircraft on an approach to land on a runway. Zero blinking lights and never heard engine noise. This craft was not in my opinion any known civilian craft I have ever seen in person. This is my 3rd sighting since September 10, 2010. and the 2nd in the last 2 nights. Last night was a bright star sized object that was under intelligent control before going dim, or out of range. (Ice blue in color) I saw it from the same vantage point I did during the V shaped sighting the night after. This thing I thought was a high-altitude plane at first, but when it slowed to a stop, I knew it wasn't a plane, then it made a curved trajectory out of sight in approximately 5 seconds. There were no blinking lights at all. It was about the size of a large star in the night sky. My sighting this evening was heading directly south, along Dufferin Road. it truly was not a civilian aircraft at all. It had no wing lights blinking, was extremely real, V shaped, with 7-9 lights underneath. It flew out of sight. The sighting lasted 45-1min seconds before the craft faded out of view heading due south from the north. This was REAL!

Hamilton
Case #69699
08-22-2015

I was walking home from work, same route as I always take, and as I turned the corner and could see the sky in front of me, there was anywhere from 10 to 15 glowing lights, similar to stars, but moving. There was a very dark storm cloud further in the background, which really helped to contrast the glowing lights, and in my opinion, they hovered in a way that would appear to be similar to someone investigating or researching the storm cloud. They floated around, and at one point almost began to join into a formation, before slowly dispersing one by one into the storm cloud.

Stoney Creek
Case #69754
08-22-15

Unexplained activity over Stoney Creek, Hamilton Ontario. Over two occasions I have witnessed strange lights happening throughout the month of August 2015. In particular on the night of August 22, 2015, around 8:00pm. I was walking the dog when I noticed two strange lights glowing in the sky coming towards my direction. At first, I dismissed the lights as planes but once the lights got closer it grabbed my attention as something else. As I stopped walking to observe the lights, coincidently the lights stopped moving as well. Once I begin walking towards the lights, they started to move away in the opposite direction from me and slowly began to fade until it eventually disappeared. I continued my walk and minutes later both lights reappeared out of nowhere and once again the two lights were at a standstill hovering in the sky. When I reached the school park a group of teenagers were there but unaware of what was happening in the sky. I approached the kids and brought their attention to the lights in the sky to at which point we were all observing the lights together. I then decided to quickly walk back home to drop the dog off and got into my vehicle and headed up the escarpment to get a broader view of the city and to see if I could catch the lights and where they were heading. It didn't take me long to find the lights as they appeared everywhere (20-30 orbs of lights now at my eye level) in the northwest direction over the Stoney Creek and Hamilton skyline. At the initial first viewing upon heading up the mountain the lights were together in a cluster hovering in the sky, not moving and what I would describe as looking like frozen fireflies until they all started to move around making all sorts of shapes and patterns in the sky. At times there were approximately 20-30 lights then it decreased to around 10 lights. At one point only three lights were visible forming into a triangle. The lights would then all reappear and at one point formed into what looked like the DNA structure. It appeared as if the lights were trying to tell a story and communicate by patterns and shapes. At one point it felt as if the lights were trying to send me a direct message. The lights were continuously disappearing and reappearing and reforming shapes for about a half an hour with me getting in and out of the car multiple times due to anxiety and excitement. At another point I observed three additional orbs of lights rising simultaneously from behind the houses to the horizon, roughly around Hamilton Harbour, and joined the other lights. The lights eventually all disappeared quickly except for one which faded away slowly. While this was all happening, I tried phoning family and friends but was having difficulty holding onto a connection to having no signal at all. A couple of days before the event I had an overwhelming feeling inside of me of some sort of presence that would make itself aware to me.

Oakville

Case #81347

03-01-2009

On March 1, 2009, I was driving through Oakville, ON at 5:45 am. I was driving down Reynolds Street headed toward the lake when I noticed what looked like a plane come towards me with its lights flashing much like a plane I thought. The lights suddenly stopped flashing and it appeared to be too low to the ground for a plane, never making a sound. It finally stopped, hovered over a short, maybe 10 story height apartment building and the area between Reynolds and Allan Street with Church Street intersecting. I had to drive under it. As I drove under it, it made no sound, and I could see 4 lights, so it appeared as though I was driving under a deformed rectangle or trapezoid. It was massive. I literally drove away as fast as I safely could. Of course, at that moment, nobody was around! The entire time I noticed it come off the lake, hover then me driving away from it probably took 5 mins or less. It didn't leave the area that I saw, I just drove away from it. At the time I was pregnant, I had a dog in the car. The dog paid no attention to it, and nothing notably happened to my vehicle.

Ajax

Case #90053

02-05-2018

I live in Ajax, Ontario just west of Harwood about 1/4 mile from north shore of Lake Ontario. Last night at about 7:35 PM I was looking south out my backyard patio door when I saw a large orange orb approaching slowing from the south southwest. This area is in the flight path from Toronto's Pearson airport which is about 35 kilometers west of Ajax. I am very familiar with the air traffic in the area which typically moves west to east (takeoff) or vice versa (landing). There were no flashing navigation lights and the object was about the size of a penny held at arms-length

with its luminosity remaining bright but constant throughout the sighting. It approached slowing from the south from over Lake Ontario in the northerly direction under the cloud cover at an altitude of about 500 to 1000 feet. I brought the object to my wife's attention, and we watched the object as it changed direction slightly and proceeded at about the same altitude and speed following the lake shore moving west to east. I asked my wife for her cell phone so I could take pictures and by the time she had given me the phone the object had moved west about a kilometer or 2 along the shoreline until it disappeared beyond my line of sight at 7:43. I managed to snap 5 or 6 pictures by which time the object was considerably smaller. When I first viewed the object, it appeared to be a single object but upon viewing the pictures the object was seemingly 2 bright disc-like objects side by side with pillars of light extending upwards similar to the shape of Christmas trees. The series of photos shows the object changing shape slightly as the glowing interior of the light seemed to be on fire. One of pictures also shows what appears to be an appendage on the bottom the very center of the disc. It is worth noting the Pickering Nuclear Plant is about 2-3 miles to the west of my residence. The entire sighting lasted only a minute or two. Regrettably, my wife's cell phone does not currently have the facility to send photos via email, but I am working to resolve so that pictures can be shared and posted ASAP.

Toronto

Case #91688

04-29-2018

Again, the object appears in same place. Airplanes are unusual in area. I previously sighted a "space plane" object in gold. It changes shape in the photo. Airplanes peculiar color seen. This is getting so common. It's as if they don't care who sees them.

Wasaga
Case #94690
09-05-2018
Witnessed a black triangle UFO hovering over the water at Wasaga Beach facing the beach and cottages. The object was stationary and silent with white lights turning on and off in a random pattern on the three tips of the triangle. The object would appear around 10:00pm EST every night for about 4-5 nights. By the morning the object was gone. Multiple witnesses. A lot of the residences were saying that it was Base Borden conducting military tests. I don't buy that. Not at this objects size and presence.

Gravenhurst
Case #95466
10-09-2018
On August 5th, 2018, 7 boaters including us tied up our cruisers on Deep Bay off the Trent Severn in Ontario. We were there for a total of 3 nights tied up together swimming, cooking, and enjoying a perfect weekend on the water. On Sunday evening my 55-year-old boyfriend and I decided to lay out a blanket on the bow of our boat and watch the stars before heading to bed. After about 20 minutes or so we noticed a large bright round object flashing red and white spinning lights across the lake and above the tree line in front of us facing north. We watched it for some time, confused why it wasn't moving and what would be flashing these types of lights. We grabbed the binoculars and took turns watching it being able to observe the colors better and confirm that the colors were turning and spinning or alternately flashing. We ran through all possibilities we could think of to explain what it could possibly be. If it was a helicopter, it was the wrong shape. How could it hover in one spot for so long, why, and how it could be spinning lights, and why was there no sound. If it was a plane, how can it hover. It clearly was not a weather balloon and far too large to

be a drone. We wanted to watch it all evening but had to get to sleep unfortunately. In the morning it was gone. I tried to take pictures, but the object was to far and the surroundings were too pitch black to capture anything but darkness. This was the oddest thing we have ever seen. When we returned to the marina the next day and mentioned something to a friend there, he also said he saw something over the water that was flashing lights and very strange. We have not seen it since.

Lake Superior, White Fish Point
Sighting date: June 12, 2019
Time: 1:00am-2:38am
Weather: overcast, some rain, temperatures lower 50's
4 Witnesses, 2 wishes to remain anonymous
Sighting description: UFO hovers over Lake Superior as freighter passes directly beneath. At approximately 1:00am, witnesses were looking for Yooperlite Rocks at Lake Superior off White Fish Point Bird Sanctuary. At approximately 1:00am the witness looked over his shoulder at the orange lights in the sky. At first, he thought they were drones but then the lights started to maneuver around the sky in an odd fashion.
Description of lights movement:
3 lights in the sky approximately 200-300 feet in altitude above the water
Distance from witness: 1-2 miles
Color: bright orange
Movement: Lights began to quickly change positions in a circular pattern, changing eventually to two lights, then one light, back to two lights and finally back to one large single light in the sky hovering.
Hovering light reflected off the water
No sound
No EMF
Only thing between light and witnesses was sand beach and water.
View thru binoculars showed light as not one large single light but several tight grouped tiny

lights in a quivering motion. Thru binoculars witness watched the light separate into two then back to one. His description was "like looking at plasma in the sky".

At approximately 1:15am a freighter (739 feet Spruce Glen, a 1983 iron ore freighter) came into the harbor by way of the channel which pasts White Fish Point and traveled directly under the hovering light. The light reflected on the deck of the ship as it passed underneath. The ship barely cleared underneath the hovering light as it passed the captains deck section.

The witness contacted the ship lines trying to recover the deck footage but was told the ship lines had nothing to report. The witness also has several minutes of footage including footage of the freighter passing directly under the hovering light.

Underwater Anomalies
Another Stonehenge at the Bottom of Lake Michigan?

In 2007, forty feet below the surface of Lake Michigan where Grand Traverse Bay Underwater Preserve is, Mark Holley, professor of Underwater Archaeology at

Figure 21 Stone Henge under Lake Michigan

NW Michigan University, found an interesting site with colleague Brian Abbot while moving across the lake in a ship that contained sonar

equipment, used to examine old shipwrecks. As they were passing thru, they found a series of stones that stimulated something very familiar, the infamous Stonehenge. The two professors first sent divers below the surface to look closer and document with underwater camera equipment. The photographs captured something very unique, a formation of long, tall stones that seemed to be placed, they were not a natural formation. Professors Mark Holley and Chris Doyle had to see the anomalies for themselves. Could this be the remnants of another 9,000-Year-Old Stonehenge at the Bottom of Lake Michigan? After diving the area Holley stated, "The whole site is spooky, in a way. When you're swimming through a long line of stones and the rest of the lakebed is featureless, it's just spooky." Holley kept the location of the Lake Michigan Stonehenge a secret to please the Grand Traverse Bay's American Indian Community and to protect the area of his research. The Grand Traverse Bay's American Indian Community wants to minimize the tourists to the site.

What appears to be a Mastodon

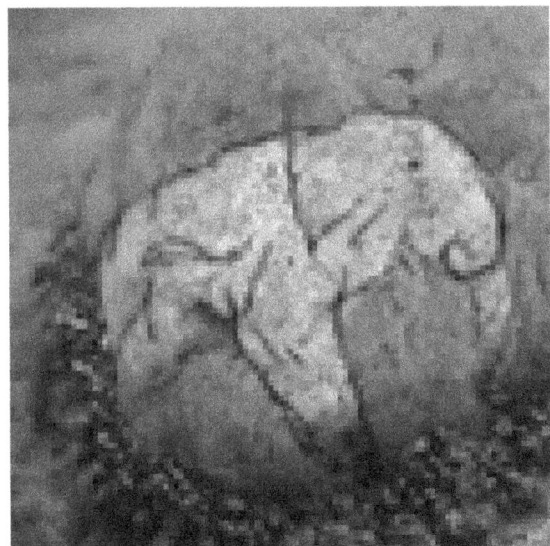

Figure 22 Mastodon carving under Lake Michigan

On a diving expedition put together to look at the stones, underwater photographer Chris

Doyle found a stone with an incredible depiction of what appeared to be a Mastodon. The carving must have been made before the Mastodons were extinct 10,000 years ago. It is made out of granite, which is a very hard material. For people to carve something onto this stone, they had to use a tool harder than granite. Mark Holley stated, "It was really spooky when we saw it in the water."

A Bimini Road in Lake Huron?

John O'Shea, a Professor at the University of Michigan and Curator of U-M's Great Lakes Division of the Museum of Anthropological Archaeology Site, made an unexpected discovery. On the Alpena-Amberley Ridge under 121 feet of water 35 miles southeast of Alpena, Michigan, appears to be a once dry land corridor connecting NE Michigan to southern Ontario. The main feature, called Drop 45 Drive Lane, is the most complex hunting structure found to date beneath The Great Lakes. The area is constructed on level limestone bedrock stone, and the lane is comprised of two parallel lines of stones leading toward a cul-de-sac formed by natural cobble pavement. Three circular hunting blinds are built into the stone lines, with additional stone alignments. Could this be a "Bimini Road" in Lake Huron? Archaeologists believe this area may have been used centuries ago, as a means for corralling caribou.

Chapter 7

Reports from the Caribbean

Cruise Ship Sightings

Figure 23 Independence of the Sea
Photo: Debbie Ziegelmeyer

Off the coast of Puerto Rico

MUFON Case #62483
January 2, 2015
The Mutual UFO Network (MUFON) received a report from a man that was taking photographs of the sunset while aboard a cruise ship in the waters off the coast of Puerto Rico. According to the report, the man was taking back-to-back exposures of the sunset at approximately 30 seconds apart. One of the photographs show what appears to be a brightly lit green object hovering just above the water. The object or objects appear to be just a few feet above the surface of the water. The green light can be seen reflected on the surface of the water. The man said that when he noticed the object on his camera's display, he looked for it to get another shot, but it was gone. The object was not visible in any of the other photographs. Case disposition: Unknown

Saint Kitts

UFO Hunter Report
Thursday August 17, 2017
Saint Kitts
While on a cruise ship returning to Ft. Lauderdale, the witness was out on his stateroom balcony when he noticed an object that first resembled a falling star. The object, coming from the NE, stopped, made several maneuvers, stopped a second time, and then moved SW right over the cruise ship at a high rate of speed. Witness description: "the object looked opaque, glowing, orb or circle."

Near Nassau

MUFON Case #97699
December 4, 2018
Royal Caribbean Mariner of the Seas
We were in blue water and a day away from docking at Nassau. I took the photo on December 4, 2018, at 5:18pm (according to my phone's documentation), from the ship's balcony. The two white images, one appears structural and the other appears to be light shining on water but not as a beam, were not in the photo when I took the photo. I noticed the images days later at home when reviewing my cruise photos. I noticed the object when I looked at my photos after my vacation. We were out on the ocean with nothing seen while taking photo other than the beautiful sunset and clouds. We thought it was a buoy, then definitely not a buoy. It appeared to be above the water. Since I didn't see it while taking the photo, I cannot state its actions or motions.

After seeing the photo, it looks very unusual to me. I felt intrigued. Asked others what they thought it could be. Some thought it was in the water, but I think it is hovering with light shining down onto the water.

Labadie, Haiti
MUFON Case #96764
November 13, 2018, 6:23 PM
Royal Caribbean Mariner of the Seas
I was on the deck of a cruise ship looking out at the city prior to departure and noticed something moving in the sky with lights on it. It moved back and forth I called for my wife, snapped a couple of photos, called for her again than it shot up into the night really fast. It moved around like it was trying to get footing in the sky like it was looking for a specific spot, then it was gone. It was dark except for the lights, a bright one on the bottom and three or four lights around the middle part. I was a bit anxious to get my wife out to see it, but by the time I went in and called her again it left, my wife never saw it.

Reports from the Coast
Cozumel, Mexico
On March 21, 2019, 17:00, I was told about a local UFO incident. I was a speaker on the 2019 MUFON cruise ship Royal Caribbean "Independence of the Sea". While on a shore excursion in Cozumel booked from our ship through Aiomar Adventours, our group was taken to a nature reserve sanctuary area where there were viewing decks across the eco-area and a large lookout tower. Naturally being a MUFON cruise and being in Cozumel on the date of the summer solstice, bonus... several of us were wearing t-shirts to the occasion. Our bus tour guide Lorenzo began to tell us of the importance of the date, with many of us smiling at the fact that he was unaware of our group and its knowledge. He finally asked, "why are so many of you wearing

Figure 24 Tour Guide Lorenzo Photo: Debbie Ziegelmeyer

UFO t-shirts? Oh, that was the million-dollar question for people on a shore excursion from a MUFON Cruise ship. Before I could respond, a lady in the group naturally asked tour guide Lorenzo if he had ever seen a UFO, and being the area we were in, the answer should not have surprised us. Lorenzo began to tell us of a sighting he had about twenty years prior in the very area we were presently in. As he described; "I was driving home along this road when I looked out to the water (as he pointed out the exact area), looking towards that small island when I saw something come out of the water and hover. It was bubble-like, large shiny metal like a mirror, with water dripping off of it." I asked him how large, and after a few examples of the size of objects, Lorenzo determined that the object was about the size of a small Volkswagen car only round. Lorenzo went on to tell us that the object followed along the shoreline, in the same direction he was traveling, but he thought was not necessarily following him. Sight of the object was lost partly with the mango tree obstruction along the road, but Lorenzo claimed, "it just disappeared". His cousin was on the same road about ¼ mile behind him and saw the object along the beach and Lorenzo's car. His cousin told him later that evening that he thought the craft was following Lorenzo. Lorenzo also told me that local fishermen see such objects and also balls of light quite frequently. He stated that "you can talk to any local fisherman, and they will tell you that they see UFOs going into and coming out of the water all the time."

Andros Island, The Bahamas

MUFON Case #89279

December 01, 2010

Was sitting on the beach with a co-worker on a sunny day and looked to the west because I heard something break surface of the ocean, nobody else was around and instantly a quick flash of light like a laser beam came out of nowhere and this beam quickly skimmed the surface of the ocean, it made a couple of water splashes, then the laser quickly took off heading south. Me and my co-worker friend couldn't believe what we had just seen, so we decided to just leave the beach.

Atlantis Resort, Bahamas

MUFON Case # 48313

May 24, 2013, 7:00PM

Just a little after the pools closed at the Atlantis resort me and my wife were getting to know the facilities when we saw a very different looking object in the sky. We were just in front of the Cove Hotel around 7pm and there weren't many people around specially because everything was closed. The object was north of us above the sea. At first it looked like an airplane breaking the speed of sound but the pictures we took are very weird. We also asked around the resort and people told us that the army bases were very far away and in the opposite direction. We filmed also but it was very dark, so I didn't attach it.

Cañón del Chicamocha, National Parks, Colombia

MUFON Case #23155

April 10, 2010

We were with four of my friends visiting some of the National Parks in Colombia called " Cañón del Chicamocha", the second largest canyon in the world, after Colorado's of

Figure 25 Beach at Cozumel Photo: Debbie Ziegelmeyer

course. We split into couples and walk different paths of the rally, then I found the best spot to take an automatic picture with my girlfriend with the canyon as a background. I set the camera on AUTO, and it shot. When I was at home checking the printed results of my trip and the events on the holiday, I was surprised when I found something unusual on the picture my girlfriend and I took on the Canyon. There was a metal oval shaped object very far from us about 6 to 10 miles, right above the canyon, an object didn't notice when we were there. I say metallic because there's a silver shine on the upper side of the UFO. It is amazing the speed it should have, because I've taken pictures of moving objects before, like aircrafts or cars, even Race Cars, over 300 Km/h or 180 mph, and I've never seen 'misty' like this one.

Pereira, Columbia

MUFON Case #52923

December 23, 2013, 5:30 PM

The witness reported seeing a powerful white light moving vertically, the light disappeared in seconds. Analysis was done by a MUFON Investigator. The appearance of the light, size, color, and behavior would correspond to ball lightning, which is still a very mysterious weather phenomenon. Data indicates that the image has a high probability of being original. Analysis: unknown

Bogota, Columbia
MUFON Case # 94949
September 1, 2018, 9:30 PM
Object was observed for more than 40 minutes in the SW sky. There was only a small movement in a circular way. The center of the lights looked totally dark. The witnesses noticed three lights blinking on the edge of the craft. On Saturday, September 1st of the current year we had friends from the university to share a pizza at the house of a companion, when we went to their home. I notified the companion of the lights I saw in the sky pointing and we all looked and realized what we were seeing. Initially it made us curious, and we thought it was a drone, but in detail we realized that it was in a static position, and we ruled out that possibility because of its appearance. We stayed more than 40 minutes observing carefully and it only had a small movement in a circular way. In the center of the lights, it looked totally dark, and we could only see that there were three lights blinking. I was personally curious, so I tried to take pictures with my cell phone and since it was not very clear in my capacity as a student of design and photography, I decided to make a photographic record of the object with my professional camera. In the approach, I managed to notice more or less the number of lights color and intensity and I managed to Zoom in on images.

Popayan, Columbia
MUFON Case # 97399
January 1, 2019, 12:27 am
The UFO ascended to the distance and passed over my house at relatively low speed and without issuing any sound, had red, yellow, and white flashing lights in the low part and

many sequential lights on the sides. The object was lost in the distance. I was in the window of my room observing the artificial fires of the new year, when I suddenly observed a light that would seem to be a star, until it started to ascend and to approach slowly and without emitting any sound. Immediately, I was excited to take my camera and I started to take photos to the unknown object, which had sequential lights of red, orange, and white colors on the sides and 3 intermittent lights of the same colors in the bottom. This body was dark, which did not allow me to see a defined form, but when I observed the photos, I can think that maybe it was square or rectangular. When it passed above my house, I ran to the opposite balcony of my room to continue taking photos. I observed the object until it was lost at distance, preserving a straight path.

San Jose, Costa Rica
MUFON Case #44737
October 17, 2012
Triangle shaped just hovered over the freeway. As I was driving home just after midnight, I noticed a bright light with flashing blue and red lights from a distance. I thought it was an airplane, but it looked like it was not moving; this from about 5-7 miles out. I figured the airplane was flying towards me and that was why it looked like it wasn't moving. As I kept driving closer, I noticed that it was just hovering over the freeway. As I was about to drive underneath it, I slowed down, opened the windows, and stuck my head out. When I looked up it was a triangle shaped craft with small white lights at each corner of the craft. It was about 75 feet up, motionless and I could hear a slight hum. I did not pull over on the highway to take a picture but as I got off the freeway and looked back, it was still hovering over the road. I drove underneath the object at about 12:30am and lost sight after I got off the freeway and onto city streets. At first, I

was in denial but after I drove directly underneath it, it turned to awe.

Punta Cana, Dominican Republic
MUFON Case #89633
January 15, 2018, 3:00 PM
Object hovered then was stationary.
The object was pointed out to me by resort staff, there were 3 objects. They were very high up, spherical, round, with no movement. They were larger than normal stars, like a Venus size and shape, but visible during the day. They were much higher than the clouds. The next day I observed an object make a 45-degree angle change with no change in velocity. It then dropped 90 degrees down, stopped, then exited the opposite direction at unbelievable speed. It was not natural and unlike anything I have ever seen. No known aircraft I have seen have those kinds of aerodynamics.

Punta Cana, Dominican Republic
MUFON Case #90317
February 12, 2018
Witness sitting poolside
I was at a resort in Punta Cana just starting my vacation lying in the sun by a swimming pool and looked up at a perfectly clear sky. I noticed an object that was moving across the sky which I thought was an airplane because of how high up it was. The object stopped and hovered for at least an hour before I lost interest in the object, so I took some video and pictures.

Puerto Rico

Puerto Rico is almost rectangular in shape, approximately 100 miles long by 35 miles wide and is the smallest and the most eastern island of the Greater Antilles (Cuba, Hispaniola, Jamaica, and Puerto Rico). In the Atlantic Ocean, Puerto Rico is one of the most documented countries for water related UFO

sightings. The Puerto Rico Trench, which is 497 miles long and 5.373 miles deep, lies along the

Figure 27 Debbie on the beach in Puerto Rico

Figure 26 Puerto Rico Photo: Debbie Ziegelmeyer

northern coast of the island and is the deepest part of the Atlantic Ocean. This could be the reason why there are many USOs (Unidentified Submerged Objects) observed entering and exiting the water by many people around the island and by tourists along the beautiful beaches. Puerto Rico is also one of the three points of the Bermuda Triangle. Also, of interest on the island of Puerto Rico is the Arecibo Radio Telescope which is the largest radio telescopes in the world.

Jorge Martin, Puerto Rico Investigative Reporter

Jorge Martin has been an investigative journalist in Puerto Rico covering the subject of UFO sightings since 1975. I interviewed Jorge Martin and was amazed at the UFO activity not only coming in and out of the ocean surrounding Puerto Rico but of craft coming out of worm holes above the island and quickly entering the water. Mr. Martin told me these worn holes open briefly, just long enough for craft to enter or depart. He also spoke of activity in the area of the rain forest in the northwestern part of the island and cylinder craft seen hovering over a nearby radio antenna. Jorge Martin began investigating the surrounding waters via NOAA and found very interesting underwater topography which could not be formed by nature. One area is just off the shores of Ponce, Puerto Rico's most populated city outside of San Juan. Another area of interest is the far northwestern corner of the island not too far from Rafael Hernandez International Airport which serves Aguadilla. A large, long narrow anomaly running along the coastline and extending north of the island were clearly visible. Jorge Martin was able to copy images of both areas before they were masked by NOAA at who he assumes to be by Government request. I searched Google Earth and was able to find the anomaly off the southern coastline near Ponce, but the northern image is not visible. Just a few days later, the second anomaly was also blocked. I asked Jorge Martin if through his 40 plus years of investigations in Puerto Rico, he believed it was possible there could be one or more underwater bases possibly occupied by alien life. Mr. Martin said it would be impossible to prove 100% because of years of government cover-up, but he believed underwater alien bases exist. Jorge Martin told me of his years of documentation of craft entering and exiting the oceans surrounding Puerto Rico and

pictures of what he believes to be a worm hole where these craft enter and exit the area. More detailed information can be found on Jorge Martin's website: Jorge Martin-Enigmas del Milenio/English version. (The Google Earth anomalies are posted on this site)

Laguna Cartagena

Figure 28 Laguna Cartagene Photo: Google Earth

Near Lajas is Laguna Cartagena, a small body of water of uncertain depth, polluted from agriculture runoff and illegal sewage drainage. The lagoon is unusual in several aspects, not the least of which is its shape: from the air, the grassy border of the lagoon looks uncannily like a map of the United States. The lagoon is also known for its large and unusual waterfowl population, some species of which are unique to the lagoon. For years, people living near the lagoon have witnessed balls of light coming and going from the water. A local professor claims to have seen craft flying into the Lagoon on many occasions from 1956 to well into the 1960s. She described the craft as being "metallic, silvery crafts with a dome at their top which was also metallic, with many windows around the dome."

Local Fisherman Sees UFO

In 1989 Inocencio Cataquet, a local fisherman and professional skin-diver and snorkeler reported an incredible sighting. "I was alone, fishing in the deep, when around 5:30pm I became aware of an enormous really shiny thing moving underwater. It stopped 500 feet away from me, and then it turned itself off. I grabbed my flippers, mask and snorkel and went in for a closer look. It was 25 ft. deep and when I went down, I realized it was round. It was flat on the bottom and curved on top. I descended even more and got right over it. I touched it and felt that its surface was not metallic but porous, like touching sand or cement and it had an ashy gray color. When I touched it, I noticed that the thing was sort of absorbing water, sucking it from the top, which was full of tiny little holes. You could feel the suction on your hand. I went up for some more air and went down again. This time I swam under it some three feet from the sand at the bottom and I felt it expelling water from underneath as if expelling water, it had sucked from the top. Suddenly it lit up and made a nasty squealing sound. The water started heating up, so I headed for the shore. After I had gotten out of the water I turned and saw this tiny little point of light rise from the water and head straight up vanishing."

Surfer Reports

A 53-year-old social worker who grew up in Puerto Rico stated: Since the age of thirteen, I have been surfing in the west coast of PR, and on many times, UFO have been sighted by me and other surfers! On one occasion, around 6.30. A.M. that is the best time to surf by the way, several silver-colored objects came out of the water about half a mile from where we were waiting to catch a wave. These objects were round and hovered in one position, then they took off straight up at a speed that I cannot explain, but they disappeared so fast it was incredible. About a half hour later, the objects came back, I think it was around 6 or 7 of them and went into the ocean. We never reported it to anybody, we just thought it was a military practice or other things that we did not know about. This all happened in the year of 1971, January, I do not remember the exact date. I am only reporting this one event, but we had many more sightings of this kind since we surfed in the west coast of Puerto Rico.

This event happened in Rincon in a beach known to surfers as Tres Palmas, and there were about 8 to 10 surfers that day.

Caja de Muertos, December 1995

The F.U.R.A. (Fast Action United Forces) an arial division of the Puerto Rico police force, encountered a huge unidentified craft near dawn in December of 1995. While driving along highway #1 in route to Ponce, a high-ranking police officer noticed a large luminous craft hovering in the sky close to the islet of Caja de Muertos. The officer at first, suspected the craft to be criminals smuggling illegal drugs. The officers first reaction was to radio in the sighting to the Police District Command Office located in nearby Ponce to investigate. Several F.U.R.A. members along with a gunner, rushed to action boarding a Hercules helicopter stationed at Mercedita Airport and took flight towards the unidentified craft. Upon entering the area, the team was confronted by a gigantic mysterious looking luminous craft. The craft made a "jump" moving forward at a high rate of speed, and almost instantly was positioned over Caja de Muertos, some distance from their current position. The unidentified craft made another "jump" to a hovering position just in front of their helicopter. Almost immediately, the helicopter pilot lost control of his aircraft which at this point was rendered "frozen" in the air. The pilot had no control over the craft which remained motionless in the air in a set position. Several minutes later, the force "ended" as quickly as it had begun, and the gigantic unidentified craft was gone in the "blink of an eye". The helicopter pilot regained complete control of his aircraft which experienced some technical problems, but the team was able to make it back to Mercedita Airport in Ponce safely. After an inspection of the Hercules, it was discovered that a turbined had burned out. The pilot and crew filed reports of the incident with the FAA and local authorities, were ordered to take psychiatric tests, and were consequently prohibited from flying for a six-month period.

Rafael Hernandez International Airport

April 25, 2013, over three minutes of video was recorded by the Customs & Border Protection crew and investigated by the "Scientific Coalition for UAP Studies" (SCU). The Video was shot from a helicopter at Rafael Hernandez International Airport. The object moves very quickly around the airport and over the water. In one part of the video, you can see how the object descends into the ocean at extreme speed. When it comes back out of the ocean, there are two objects on the screen.

The UFO incident began at approximately 9:20 pm April 25, 2013, at the Rafael Hernandez Airport in Aguadilla, Puerto Rico. The unknown craft was seen and filmed by the crew of a DHC-8 Turboprop aircraft from U.S. Customs and Border Protection (CBP), a division of DHS. Having just taken off from a routine flight, a crew member noticed "a pinkish to reddish light over the ocean that was in their vicinity and approaching toward the south." Believing this to be incoming traffic unreported to them by the tower, a call was made to the tower for the updated information. The tower informed them that they had the craft in sight and were trying to determine its identity. As it approached the airport it's lights dimmed and turned off rather than brighten for landing. DHC-8 immediately began tracking, circling, and filming the object using their onboard thermal imaging system, keeping their distance as they filmed. The object was not visible on the DHC-8 radar mainly because their on-board radar was positioned downward to track ocean vessels. Airport

records indicated that a FedEx flight on schedule to depart at 9:10pm was delayed until 9:26pm, possibly due to the incident at hand.

It was determined from the recorded flight plan that the DHC-8 took off at approximately 9:16 pm and circled the airport twice before leaving the area about 10 minutes later. The video released by the Department of Homeland Security (DHS) is just under 4 minutes long. The unknown craft was tracked by DHC-8 for approximately 2 ½ of these recorded minutes.

The research group "Scientific Coalition for UAP Studies" (SCU) consulted thermal imaging experts to determine the capabilities of the system used for the filming. The group also analyzed the video frame by frame in an effort to determine the type of craft, approximate size, speed, temperature, and its flight path. The group was able to determine that the object traveled in from the ocean, from the north or northwest of the airport's airstrip, flew over the airstrip, then turned back to the north and headed back out into the ocean. It had moved along the shore, turning to the south, and eventually was lost sight of over the ocean. The research group's report states that: "The object was between three to five feet in length and its speed varied between approximately 40 mph to 120 mph. Its median speed was roughly 80 mph. Its speed through the water reached a high of 95 mph and average 82.8 mph. Notable characteristics at the end of the flight was when it apparently submerged into the ocean, traveled for over half a mile, and then flew back out." At this point the craft either joined with a second craft or divided into two separate craft as seen and recorded on the DHC-8's thermo imaging video.

The SCU research group through extensive research, was able to rule out the unknown craft as being a balloon due to the current weather conditions and wind speed, and that the object was traveling too fast to be a bird. They discovered that peregrine falcons, which do occasionally visit Puerto Rico, only fly at an average horizontal speed of 40 to 56 mph, and a maximum of 65 to 69 mph. The craft being investigated was estimated to be traveling much faster. The group also looked at the possibility of the craft being a drone. The Navy is working on a drone called the "Flimmer", that can fly through the air then dive into the ocean and become a submarine, but current Flimmer drones have not been tested underwater and only have an airspeed of 68 mph. They also noted that even the fastest known underwater battery powered torpedo only travels at 50 mph.

The final conclusion determined by the SCU was as stated: "There is no explanation for an object capable of traveling under water at over 90 mph with minimal impact as it enters the water, through the air at 120 mph at low altitude through a residential area without navigational lights, and finally to be capable of splitting into two separate objects. No bird, no balloon, no aircraft, and no known drones have that capability." The group stated the following: "The source of this video evidence was vetted and identified, and they are absolutely certain that the information received comes from sources on board the DHS aircraft and the video is genuine. Their investigation also confirmed what they were told by their sources.

Culebra Island

MUFON July 8, 2014
The incident took place last year on the island of Culebra, 17 miles east of the
Puerto Rican mainland. My boyfriend and I had been living on the little island of Culebra PR for the past 3 and a half years. Towards the end of

that time, we moved out to the end of the island where the skies are very dark, and the property we lived on was fairly remote at the furthest eastern end of the island. We had been out on the front deck of our house enjoying the starry skies when we noticed a fairly large light to the north end of our porch. It was completely visible to us, as it was quite high in the sky. We watched it for quite some time, and as we were so relaxed, we figured it must be a plane heading to one of the larger islands. We were very familiar with the traffic patterns of the airlines by this time, and we thought that the object was kind of off course and at somewhat of a strange angle, even the time of the flight seemed a little off schedule (11:40 pm approx.) but we were talking and enjoying the night. We continued to observe the object, but by this time we realized that this "flight" was not really moving. It seemed to stay in one place, the more we looked at it, we noticed that it had kind of a hazy field of light around it and what seemed to be like a beam of light coming from it. The angle of the light was very odd, not like a plane at all. It was not scanning the area or moving, just hovering. The light of the object itself began to pulse variably from white haze to a green haze and there was a strange kind of static feeling and sound in the air which would not be normal for tropical weather. It was almost kind of hypnotizing. The object hung in this position for somewhere between 20 to 30 minutes, but then we realized that it seemed like it was slowly getting closer. We both laughed and said it felt like they were looking for something, or maybe it was even us they were watching. I definitely got and eerie feeling at that point. I went inside for a minute to use the bathroom and when I came out it was still there, just seemed closer. We watched it for approximately another 10 minutes or so and then suddenly the glowing object made a rapid 90-degree dodge to the right (east) and vanished instantaneously.

MUFON Sighting Reports Puerto Rico

MUFON Case December 20, 2014
12:00 AM: I was travelling with my wife and my 3 children in a mountain road in Cayey (Puerto Rico), when I saw two orange lights in the sky. I stop my car and one of the lights wasn't there and the other turned off the light, and I saw the object, it was like a black lemon. Then the object began to descend into a mountain, and I lost it.

MUFON Case #76274
April 27, 2016
A former teammate of work in INTERPOL Puerto Rico sent MUFON (2) photos taken by his sister with a cell phone while passing the area off the northwest coast of Puerto Rico. The witness reported that when taking pictures of the coastal area, the images on them were not visible. Only seconds after shooting, she checked them observing two (2) images of UFO's. The pictures were sent to Robert Powell, Director of Research for MUFON, for analysis.
Case disposition: unknown

MUFON Case #80879
October 01, 2016
"The object seems to be saucer shaped. It is tilted relative to the vertical axis by approximately 45 degrees. In the image, the left side is seen lower. The object appears to have on the top a very bright dome in the middle. The light of the dome appears to emanate from it, and not just reflected from the sun. The bottom of the object is observed to be very dark. Of particular interest is the observed concentric halo around the object.

January 2006 Personal Sighting

On January 25, 2006, my husband and I were on a Royal Caribbean cruise sailing on Explorer of the Sea. At 6:20pm shortly after we had left port from St. Maarten at 6:00pm, I was taking pictures from the right side of the ship thru an open window towards the shoreline. We left St Maarten and had passed all of the area islands, sailing out to sea headed towards St Thomas where we would dock Jan 26th. I was taking pictures looking north just passing the last Island when I noticed something odd in the water out from the ship. I remember turning around and looking to see what on the ship could glow like that on the water, but these objects were moving and there was nothing lit behind me.

At first, I thought it was a mini sub or frisbees thrown from the ship, but these objects didn't seem to be connected. As we moved passed the two objects which stayed stationary in the water, I took one quick picture and then ran towards the back of the ship trying to get a second shot which I did. I didn't have time for a third picture before the objects would disappear behind the ship in the distance. I spoke to the ship Steward, asking about mini submarines in the area or Frisbees being thrown in the water like skipping rocks. The steward informed me that no mini subs came out this far and it was illegal to throw anything from the ship. The sun was SW 12 degrees on the opposite side of the ship and being daylight, I couldn't locate any lights nearby on the ship that would reflect on water or even on a window!

I sailed twice on Explorer of the Sea, this was the second time, and have sailed on nine cruises to date. I take a lot of pictures and have never been able to duplicate this. After I this incident while still on the ship, my camera clock setting changed from January 25th to February 5th, 11 days and 7 hours later than the correct time and date. The second picture, which I believe I took through a closed window as I was running towards the back of the ship, has picture data for February 6th with several nighttime pictures taken hours later in between the first shot and the second shot. I also noticed that I not only captured 2 objects in the water, but a third object is in the sky on the top left side of the picture. The second shot I took only captured one object in the water.

Figure 29 Unknown objects under the sea
Photo: Debbie Ziegelmeyer

Figure 31 Close up of one object

Figure 30 Close up of two objects in the water and one in the sky

Puerto Rico May Build UFO Airport

LAJAS, Puerto Rico Mayor Marcos Irizarry's is supporting the building of a UFO landing strip to welcome extraterrestrials that are frequently seen over the area. A green sign in southwestern Puerto Rico proudly displays a silhouette of a flying saucer:
"Ruta Extraterrestre." The construction cost for Route 303 will be a $100,000 project.

Directions to sign: Ruta Extraterrestre PR 303 Lajas, Puerto Rico From Ponce, take road #2 towards Mayaguez. After Yauco Plaza, take the next exit towards state road 116 and the city of Guánica. Follow the road and take road 305, which connects to road 303.

Figure 32

Chapter 8

Military, Cargo and Fishing Vessels

The Okhotsk 1908

On October 22, 1908, 11:00 p.m. while in the Sea of Okhotsk, Russian Steamship Okhotsk came across a strange anomaly. Russian Imperial Navy doctor F.D. Derbek who was aboard, reported seeing a bright greenish-white luminescence appear under the ship's stern. Quickly expanding, the luminescence surrounded the ship taking on an oval shape. The light moved with the ship for some time before separating, moving to the side then forward ahead of the vessel. Two to three minutes later the oval light reached the horizon where it brightly appeared as a clear streak which reflected on the distant clouds. Several Soviet and Russian magazines carried articles of this strange encounter, and in years to follow, many additional stories appeared of strange unidentified luminescent "wheels" crossing the ocean from horizon to horizon.

The Maria 1936

A 1936 UFO was seen and documented from a Russian transport steam vessel The Maria by shipmates engineer Lev Popov, Felix Zigel and Anton Anfalov. The Maria had set sail from Leningrad to cross the Pacific Ocean and was well underway when the incident occurred. Early one morning the ship captain noticed a strange vessel ahead of them emerging from the waters below. The ship captain attempted to communicate with the unknown vessel by radio signal, but no communication was achieved. The crew monitored the vessel, which was dead in the water, for several moments, but there was no sign of activity. It

was decided by the captain, to send a small boat from The Maria with five sailors aboard, to investigate. Reluctantly, the five sailors rowed the boat towards the strange unknown vessel as the rest of the crew watched from the deck of The Maria. But, before the sailors had reached their target, the strange vessel began to move away, up into the sky at a high rate of speed, vanishing into the clouds, with the five sailors vanishing as well. Several hours were spent trying to locate the strange vessel and the five missing crew members but neither were to be found. The ship continued to set sail on route to its destination arriving safely.

On the return trip across the Pacific Ocean, the Captain of The Maria decided to retrace his path back, including the slight detour, in hopes of retrieving his five missing crew members. Near the point of the previous location where the encounter first occurred, the strange vessel was seen again and this time, much closer. The strange vessel was something never seen before by the Captain of the Maria or its crew. It appeared to be made from a "transparent material" and inside they could see three of the five missing crew members. When the Maria approached to reclaim their three crew members, one of the sailors began to scream hysterically jumping from the strange vessel into the water. The sailor's body did not resurface, and the remaining two sailors were safely taken back on board The Maria.

The two recovered sailors told the crew of The Maria of their frightening experience. As they had neared the strange vessel in their rowboat, they found themselves suddenly on board and inside. They claimed to have encountered "strange entities" in "dark shiny suits" that looked like frogmen (scuba divers) but still "human-like". These strange entities spoke to the crew not with words but directly into their minds. According to the two sailors, these strange entities informed them that "there was no way back" and that they should stay there in this strange vessel with them. Two of the sailors agreed and were led away. The two recovered sailors had no idea where their two shipmates had gone nor what had happened to them. The three sailors who had not been led away were taken into an "isolated room" where they estimated they were kept for two days. The next thing the sailors remembered was being inside the transparent boat floating on the waters of the Pacific Ocean. They did recall though, that inside the vessel with them was a supply of strange tablets in transparent packets which they described as being their food. The entities had also offered the sailors these packets at an earlier point while on the vessel. According to the sailors, one tablet would be all one would need to eat per day for nourishment. The two recovered sailors had taken one of the tablets, but the sailor who jumped from the boat had refused. The two sailors having eaten the tablet, felt nourished and were able remain calm and eventually sleep, but the third sailor who refused to eat, began to mentally deteriorate throwing all of the given tablets overboard.

Not too long after the remaining two sailors were back on board The Maria, strange waves began to circulate around the transparent vessel. The waves would surround the vessel, eventually pulling it down under the water. Upon their return home, the captain of The Maria filed a report with the Ministry of their experience while at sea. Both he and the two

remaining sailors were immediately summoned to appear before their superiors for interrogation. The captain and sailors keep steadfast to their story of their ships encounter and experiences but were eventually silenced. The two sailors were reassigned to other ships and the Captain of The Maria was transferred to another, much lower key vessel.

Soviet Battlecruiser 1950s

Professor Korsakov from Odessa University recalled a conversation with a Soviet Navy officer, who was a friend. The officer back in the 1950s while stationed on a battlecruiser, was approached from behind by a UFO surfacing from under the waters of the Black Sea. He supposedly gave a photograph of the object to Korsakov which is said to still be in his possession.

Trindade Island

Magazine article: "Unopened Files", Number 11; Summer 1999

"Shortly after midday on 16 January 1958, a series of photographs were taken from a ship anchored off Trindade Island, about 650 miles from the coast of Brazil. The photographer, a Brazilian named Almiro Barauna, claimed to have seen a dark grey 'object' approach the island, fly behind a mountain peak, and then turn around and head back the way it came, disappearing at high speed over the horizon. The object glittered and was surrounded by a green mist, and it displayed an undulating motion, changing to a 'tilted position' as it passed over the island. On board the ship with Barauna were some 300 other crews, and around 50 of them are claimed to have seen the object." The case is well documented, and the reader is referred to several good reference books for a full account of events.

Anapa Area

A regular shark-hunter in the Black Sea known as B. Borovikov, told of an encounter with what appeared to be an underwater "alien lifeform". Borovikov was diving in the Anapa area, when he witnessed several huge creatures heading towards him about eight meters away swimming up from below. He described the creatures as having "fish-tales", were "milky-white" and seemed to have "humanoid faces". When one of these creatures stopped to stare at him, he noticed that it had "giant bulging eyes". Lieutenant-Commander Oleg Sokolov is one of the few high-ranking Soviet officers to publicly speak of unidentified fast moving "submarine" encounters. He spoke of vessels picked up and logged by sonar, traveling faster than anything known to technically exist. Technicians onboard Soviet submarines would regularly receive "strange targets" on sonar which resulted in underwater chases with the chased being faster than any known underwater vessel, performing unmatchable maneuvers and escaping with ease. Sokolov told several of his students that he witnessed an "underwater take-off" while on board a submarine in the early 1960s.

Black Submarine With "Little Men"

(From the Archives of the Malta UFO Research Center)

20 miles S of Malta; witnessed by many

A black submarine, or other type of craft traveling with no sound.

Fishermen on a boat 20 miles south of Malta were raising their nets with a catch of fish when they saw an object floating on the water's surface that looked like a black submarine. The fishermen were frightened because they thought it looked more like a monster than a submarine, so they quickly pulled in their nets and started the boat's engine. At that moment a bright light from the "submarine" lit up the whole area and "little men" began running over the deck of the object. The fishermen couldn't make out much detail from their boat but when the light illuminated the "little men," they could see some sort of apparatus around their waist. When the witness was asked how tall these men were, he replied, "About the size of a ten-year-old boy." After a few minutes, the "little men" entered the "submarine" which began to glow so brightly that the fishermen couldn't see the object. It then submerged. reference: mufor.org now defunct

"Kvakeri"

In the 1960s and 1970s, Soviet nuclear-powered submarines reported encountering strange sounds coming from a very deep depth from unidentified moving objects. The sounds were described as "frogs croaking and were nick-named "Kvakeri." "Kvakat" translated from Russian means "to croak", which is probably where "Kvakeri" originated. These strange sounds were also reported to be heard by Soviet ice-breaker vessels. It was reported that on occasion, these "Kvakeri" would circle the submarines, changing the frequency and tone of the "signals" as if they were trying to communicate. This activity would continue until the submarine would leave the area of the encounter.

The RADUGA

In August of 1965, while navigating in the Red Sea, the crew of the steamship RADUGA, observed a mysterious craft. Approximately two miles from their ship, a fiery sphere-shaped craft shot out from the water and hovered over the surface of the sea, illuminating it. The sphere-shaped craft was sixty meters in diameter, and it hovered above the sea at an altitude of approximately 150 meters. Moments later, a very large pillar of water rose up as the sphere-shaped craft emerged from the sea, collapsing just moments later.

Soviet Submarine 1965

In 1965, the crew of a Soviet nuclear submarine while in the Atlantic Ocean waiting to rendezvous with a ship, observed a UFO. The submarine having arrived and hour and a half early, was surfaced with crew on deck under a starry cloudless night. The watchman reported observing a cigar-shaped object moving silently across the sky undetectable by radar. Then suddenly, three rays of light shot out from the UFO. The craft was described as not having any gondolas, no horizontal or vertical rudders, and was approximately 200-250 meters long, not an American dirigible as first thought. The UFO then descended to the surface of the ocean about a half mile in the distance with the three beams of light still shining as though they were some type of spotlights. The craft then descended underwater emitting a strange sound which was picked up on the submarine's sonar. The sound resembled a "hissing" sound but was only detectable for a very short time.

The Charvak Reservoir 1975

In Soviet Uzbekistan in July of 1975 at approximately 3:00 a.m., four young people who were vacationing on the shores of the Charvak Reservoir, were awaken by a bright light described as being "hundreds of times brighter than the light of a day lamp". A luminescent sphere ascended from the water approximately 2,600 feet off the shore forming concentric circles around it which were different thickness and brightness. It then slowly and silently, moved across the lake being in view for approximately seven minutes.

The Severyanka 1976

Vladimir Georgiyevich Ajaja, who served as the head of an underwater exploration expedition aboard the Soviet submarine Severyanka, sighted a very "strange" creature during one of their dives. Although his report of this incident got him fired from his then position, Ajaja is reported to have had a meeting on November 17, 1976, with the Department of the Underwater Research of the Oceanographic Commission of the USSR Academy of Sciences. Among the present were the Chairman of the Department P. Borovikov, his deputy, E. Kukharkov, and twenty-nine other officials. The topic of the meeting was a lecture by Ajaja, of UFOs and the association of such, of underwater issues. Ajaja, at the time, was a duty of Chairman Borovikov's. After the lecture, it was decided to include a collection of UFO sightings over bodies of water and the depths of the hydro sphere, into scheduled activities of the Department of Underwater Research for further analysis.

The VASILY KISELEV

In December of 1977, the crew of the fishing trawler VASILY KISELEV observed something unusual and unidentified. They were not far from the Novy Georgy Island when a strange craft rose vertically from under the water. The craft was doughnut-shaped and approximately 300 to 500 meters in diameter. It stopped and hovered overhead at approximately four to five kilometers in altitude which immediately rendered the trawler's radar station inoperative. The craft remained hovered in place for three hours, before it suddenly disappeared.

Barents Sea 1977

On October 7, 1977, nine shining disks visited a floating service base for Soviet submarines in the Barents Sea. The disks descended and circled the base cutting off all radio communication in or out for eighteen minutes until the disks left the area disappearing into the distance.

Western Dvina 1979-1980

Between 1979-1980, UFOs had visited a Soviet submarine base at the Western Dvina every week during a six-month period. Disk shaped craft hovered over weapons sights, and classified areas undetectable by radar with photographing of the craft always being unsuccessful due to overexposure. In the mist of these events, sailors were told the craft were known military exercises and tests. A criminal case was opened when four pigs had disappeared from an open-air cage. Sailors stated that a disk-like object hovered over the cage, emitted a blue ray, and the pigs were lifted from the cage and onto the craft above. This incident was confirmed by the warrant officer responsible for the cage. Area sailors were accused of stealing the pigs but eventually the criminal case was closed with the sailors being found innocent. No trace of the pigs was ever recovered.

Soviet Project 671 Submarine 1979-1980

During the winter of 1979-1980, the crew of a Soviet Project 671 submarine encountered a UFO. Commander Aleksey Korzhev reported seeing a silvery disk hovering fifty meters above the surfaced submarine as it came into the navy base. The UFO was reported by the crew as keeping pace staying a bit ahead of them. It then emitted a pillar of bright white light which did not quite reach the water surface but slowly descended. Commander Korzhev immediately ordered a change of course and the disk slowly ascended disappearing into the clouds. It was speculated that the disk was scanning their newly installed up to date weapons system.

"Paleocontact"

Appearing within the pages of recently released Soviet files, the subject of paleocontact is discussed. Paleocontact, is a Russian term associated with the "ancient astronaut theory". It refers to an ancient account from the Mediterranean Sea of people witnessing a "strange underwater vehicle surfacing at high speed".

USS Joseph Strauss

USS Joseph Strauss DDG-16 Photo: public domain

MUFON #43820
1980-09-18 12:00AM
A sailor who was stationed aboard the guided missile destroyer USS Joseph Strauss home port Pearl Harbor HI, filed the following report with MUFON. While deployed to the 7th Fleet in the Far East, they participated in an operation called "Free Seas '80" (operational code name of Dakota Run), a "show of force" exercise. The exercise was an electronic intelligence collection mission across the Sea of Okhotsk as a demonstration of "freedom of the seas". The Russians considered this area an inland sea.

The sailor reported the following "best recollection of the incident" which occurred between September 14th and 24th 1980. During the exercise and witnessed by Bridge, CIC (Combat Information Center) and Signal Bridge personnel, the USS Joseph Strauss was part of a four-ship task force on a northerly course in a diamond formation. The Strauss was at the bottom of the formation and approaching potential hostile waters. The ship was at "condition 3", a heightened state of alert that included battle ready stations, but below "GQ". Under strict EMCON (Emissions Control) conditions the Strauss was the only ship with surface radar active. During the transit north from Japan to the Sea of Okhotsk, the sailor was awoken around 2 AM local time and informed that the intelligence collection team had been called away. When the sailor reached CIC, he was informed that they had a "pop-up radar contact" approximately 2,000 yards off the port beam. Under "standing orders" for a close contact, the ship Captain had also been awakened. A pop-up radar contact is defined as "something that appears suddenly" close to the ship on the surface and assumed to be a submarine. No prior confirmation was given the intel team of a submarine being in the area. The sailor checked the surface radar and noted that a plot had been started manually tracking the "contact" in relation to the ship's course. He directed an electronic search of all bands by EW and a passive sonar search, both of which were negative. A negative search result was not unusual, but no sonar emissions, even with sonar trained on a direct bearing was.

The sailor left CIC and went to the signal bridge where several people had already gathered near the signalman on duty who was looking through high-power stationary binoculars at the "contact". The sky was pitch dark, overcast, a moonless night, and all that could be seen were two white lights. At first the lights were assumed to be running lights in what could be a bow and mast running light which would be consistent with a submarine. When the sailor looked through the binoculars, he could not make out any shape or wake either from the bow or stern. He decided to return to CIC, passing through the bridge, to

report to the captain what he had seen. He informed the captain that they had no radar or sonar reports, but the object appeared to be a submarine based on the observation of the running lights. Checking the objects plotted path, he noticed that the "contact" was tracking with the ship, but moving slightly closer, then moving out a little and repeating this with no discernible pattern. The "contact" stayed with the ship for approximately 20 minutes before things began to change.

"Surface" reported the "contact" had increased speed and was now moving ahead of them. Reports from the signal bridge came as a surprise, the "contact" appeared to be gaining altitude. The sailor rushed to the signal bridge in time to see the "contact" slowly pulling ahead of their ship and ascending. The sailor stated that the object "was not going very fast but was clearly moving up and away from them". He still had the two white lights in sight, but no stern light and the object was clearly above the horizon. The sailor returned to CIC asking the TAO (Tactical Action Officer) for permission to light off the air search radar. If the object continued to gain altitude, they would lose contact on the surface radar. The TAO didn't have the authority under the current EMCON (Emissions Control) condition and radioed AB (Able Seaman) permission was denied so contact on the surface radar was lost. Permission to light off the radar was also denied. The TAO reporting a "flying submarine" to AB probably had a lot to do with the lack of response and permissions for additional radar. The uncooperative crew members involved most likely thought the sighting was a joke, and the captain certainly wasn't happy with the outcome and comments. The sailor stated that "manual tracking" was made but was thrown away at the end of the incident, so no physical evidence exists to this sighting and unfortunately, no written reports were made.

Kola Peninsula 1980s

In the early 1980s, a similar sighting was reported in the same area of the sea.

In an article written by Grigory Tel'nov titled NLO, which appeared in the newspaper Zhizn, Tel'nov told of large numbers of unidentified underwater objects sighted in the 1980s in the northern seas of the Soviet Union. Apparently Soviet Ufologists analyzed reports from various sources, made comparisons, and concluded that residents of the Kola Peninsula had seen and reported at least 36 different UFO sightings between 1980-1981. These reports were of unknown, unidentified craft ascending from the waters nearby.

The Nikolai Boshnyak 1982

In August of 1982, N.S. Krokhmalev while aboard the vessel Nikolai Boshnyak near Shikotan Island, the watch commander, and some crew members, observed an elliptical circle with clearly outlined borders which appeared on the water surface around the ship. At the time, the ship was heading towards the Yekaterina Strait between the islands of Iturup and Kunashir. The circle's luminescence was "like that produced by a TV set", traveling silent, and was estimated to be 125 meters long and 74 meters wide. An additional strange circle was seen around the moon and a third circle could be seen 3-4 meters below the water surface. The unknown seemingly surfaced circle followed along with the Soviet vessel for two hours before suddenly disappearing. None of these circles were picked up on the ship's instruments.

Andros Island 1983

MUFON Case #62741
January 3, 1983, 13:00 Hours

I saw a domed disc 1000 ft in diameter X 50 feet depth at the apex of dome.

I was aboard a military C-12 enroot from NAS JAX to NAVSTA Roosy RDS at 13hrs 3 January 1983. I was a Civilian employee of COMNAVAIRLANT. First craft was at about 1000ft headed NW, speed about 300knots. It was a white domed disc, no appendages no fin or wings. A second craft appeared at 1310hrs headed NE. The speed and discretion were the same as the first craft. We were in the area of Andros Island At the time of the sighting, I was 45 years old and a civilian military contractor flying on a C-12 military aircraft from Jacksonville, FL to a base on Puerto Rico. The other passengers were the pilots and 11 military personal who were sleeping. The pilots would have seen it, but he decided not to ask them once they landed in Puerto Rico because a pilot buddy of his informed him earlier that if a pilot admitted to seeing a UFO that they would be immediately taken to a psychiatric ward. The other witness was employed by the military after the sighting to work in nuclear missile operations. He was in the Navy and trained to be a machinist then went into the civil service.

Three days later.......

An unidentified object was seen off the stern of a Navy ship in the same vicinity.

At dusk, there was still light in the sky, so the crew were dead in the water for Navy exercises. White is their stationary position they saw 6 white orbs that seemed to be attached to something. They could not see structure even though the object was only 20ft away right over the fantail rails not 10 feet over the water. The lights were bright but not blinding, approximately the size of a large tractor-trailer tire. The craft was at least 20ft in length, no sound and completely still. After 30 seconds it shut its lights off in sequence left to right and just disappeared right before their eyes

The GORI

June of 1984

An unidentified craft was seen by two different vessels in the Mediterranean Sea. The GORI was in the Mediterranean, twenty nautical miles from the Strait of Gibraltar. At 16:00, sailor Globa was on duty with Second-in-Command S. Bolotov. As they were standing watch at the left bridge extension wing, astern, the sailors observed a strange unidentified craft. The unidentified craft looked to be gleaming, had a grayish metallic shine, its lower portion was round, and the diameter about twenty meters. As the craft stopped suddenly, it caused waves on the outside plating of the ship. Globa described the base of the craft as having "two semi-discs, the smaller being on top; they slowly revolved in opposing directions". Around the edge of the lower disk, Globa saw "numerous shining, bright, bead-like lights". The lower portion of the craft looked "even and smooth", the color of an egg yolk, with the middle having a "round, nucleus-like stain". There was something that resembled a pipe on the edge of the craft's bottom that glowed with an "unnaturally bright rosy color, like a neon lamp." Globa went on to describe the craft as having a "triangular-shaped something" attached to the top of the upper disk. The object at the top of the disk seemed to move in the same direction as the lower disc, but at a much slower pace. The craft suddenly jumped up several times, as if it was moved by some type of invisible wave and Globa could see many lights illuminating its bottom portion. The crew of GORI used a signal projector in an attempt to attract the craft's attention. Many of the sailors were on deck observing this strange craft including Captain Sokolovky and his Second-in-Command. As an Arab dry cargo ship approached the GORI, the unidentified craft focused its attention towards the other vessel. The Arab ship was on its way to Greece and confirmed that the unidentified object had

also hovered over their ship briefly. Just about a minute and a half later, the craft changed its flight's trajectory, listed to the right, picked up speed and ascended rapidly. The craft was seen again in and out of the clouds due to sun reflection but eventually flared up and was gone.

The Akademik Aleksye Krilov

On August 28, 1989, in the area of Kola Peninsula, as reported in the Soviet Navy newspaper Flag Rodini, a UFO was observed by the scientific research vessel Akademik Aleksye Krilov. The research vessel was approximately five miles out when its crew observed a luminescent cloud. Inside the cloud there seemed to be something flickering as the cloud moved across the sky at about the speed of an airplane. The flickering object suddenly parted from the UFO traveling quite a distance before disappearing. The remaining part of the UFO exploded emitting some type of gas, followed a few seconds later by an additional explosion.

The Volgoneft-161 1989

On August 2, 1989, as reported in the Tikhookeanskaya Gazeta newspaper, Captain O.I. Zimakov of the Soviet tanker the Volgoneft-161, spotted a UFO while at sea, at about 35 degrees in the northern sky. Zimakov described the craft as a pale yellowish sphere with a strange luminescence surrounding it. Several sailors aboard witnessed as the unidentified craft ascending as it moved northeast towards the horizon, disappear after about five minutes.

The Mikhail Lomonosov

Academician Shnyuyukov while on an expedition in the early 1990s in the Black Sea, told of a strange underwater unidentified vessel. Shnyuyukov was aboard the scientific research vessel Mikhail Lomonosov when an extremely large unidentified object was detected at a depth of between 1400-1800 meters. It was elliptic in shape and appeared to be approximately 6,550 feet by 9,800 feet in size. The object registered on sonar as a dense cloud up to 270 meters thick, but no hydrochemical anomalies were found in the area. Another odd feature of the unidentified object was that the Russian devices that safeguarded barometers against soil impact, began functioning while in the vicinity of the object.

NATO 1993

During a powerful storm on February 6, 1993, a NATO squadron approaching three American destroyers, were sent a radio signal by the destroyers to stand down at a distance of three miles. As the NATO squadron took heed of the warning laying adrift, sixteen bright amber flying craft appeared over the American destroyers. The unknown craft hovered for a few minutes then departed at a high rate of speed. A few weeks later it was reported that one of the American destroyers had disappeared and a search was conducted by Russian and NATO forces. A Russian ship reported registering an underwater unidentified object traveling at 60 knots on April 15th measuring 210 by 120 meters. The American destroyer though, was never found.

The Yakov Smirnitsky 1995

As reported by the newspaper "Anomaliya", Scientist Alexey Blinov, who headed an Artic expedition, and another unnamed scientist, witnessed an incident on August 13, 1995. At 8:00p.m., Blinov, while aboard the ship Yakov Smirnitsky, observed an underwater bright object approximately 100 meters in the distance. It traveled perpendicularly with the ship and was described as being a "bright round spot", fluorescent, and approximately three meters in diameter. Its speed was described as even and slow, and as it approached the ship, moved down under the bottom of the vessel, ascending again fifteen seconds later on the other side of the ship.

Blinov stated that what he saw was of "non-natural" origin.

Navy Black Sea Fleet

Published in the magazine NLO in 1997 and authored by Nikolai Sadkov, was Sadkov's personal account of an incident while he served in the Black Sea Fleet. While aboard a navy boat with other crew members recovering unspent Soviet torpedoes from a practice exercise, a strange object suddenly appeared out of the clouds above. At the time they were recovering a new secret Dolphin-like torpedo. The "spaceship" was bell-shaped, 15-20 meters in length, and descended slowly, hovering directly over the torpedoes at approximately five meters. The crew heard a voice from the sky speak in a clear Russian dialect telling them that "nothing bad would be done to them, and that everyone must remain where they are now." A round platform extended from the bottom part of the "spaceship" and attached the torpedo like that of a magnet. The ship's sonar operator came running out with camera in hand intent on capturing the event as it unfolded. Suddenly, a thin brightly red ray emitted from the "spaceship" touching the sonar operator's head, causing him to fall down. The strange voice calmly but in a demanding manor, once again told the sailors to "remain calm where they are, for nothing bad would happen to them. The "Spaceship" then flew away at a high rate of speed taking the torpedo with it. After two hours, the "spaceship" reappeared over the deck of the recovery boat, the bottom of the bell-shaped craft opened and the taken torpedo was slowly returned being placed on the deck of their boat. The "spaceship" then just disappeared. After the incident and interview by navy officials which followed, the crew was given a non-disclosure statement to sign.

The Caspian Sea 1997

It was reported by Perekrestok Kentavra newspaper that on July 21, 1997, at 10:20 p.m. Moscow time, an unidentified flying object crashed into the waters of the Caspian Sea. A message reporting the incident was immediately sent to the marine rescue services by the captain of the tugboat Schukin. The tugboat crew described the object as shaped like a helicopter, dark in color, coming from the direction of the shore. As the object made its way out to sea, it sharply changed its trajectory and crashed causing an explosion followed by a giant pillar of water. No military or civilian aircraft were in the air at the time of the crash, and none were reported missing. Oddly, captains of area ships were immediately ordered to 45 degrees 03 minutes north latitude and 48 degrees 00 minutes east longitude to conduct a search. No trace of anything was found by the ships or boats. Depths in the area waters reached 66 plus feet with stormy weather, so divers were not deployed. Air Traffic Controllers from Makhachkaka Airport reported that no aircraft under their control were missing, and the Soviet navy denied any on-going tests of secret aircraft in the Caspian Sea area.

U. S. Virgin Islands

May 12, 1998
MUFON Case # 31972
May 12, 1998, 20:00 Hours
Serving on board a U.S. Navy Destroyer, out for Naval missile and bombing war exercises.
Noticed a formation of about 6 white orb-like lights following ship.
Suddenly one light went off, then the next one, then the rest in order.
At the time if the incident(s) The witness was serving on board a U.S. Navy Destroyer and was sent out to sea around Puerto Rico and the U.S. Virgin Islands in the Bahamas for Naval missile and bombing war exercises.

The first incident took place during an UNREP (taking on fuel, which means pulling along-side a tanker ship while still in motion). It was a night of a UNREP, the witness was assigned with a few other shipmates on the flight deck, getting ready to receive a distance line that the crew from the tanker was about to shoot over to the ship. The witness was looking around at the tanker and noticed that their mast had an unusual large number of lights, as he continued to look at the mast, he noticed a formation of about 6 white orb like lights had separated from the tankers mast lights. The witness realized that it was an aircraft following them. The lights seemed to be in an arch-like formation from his point of view and seemed very interested in what they were doing. Looking around the deck of the ship to see if anyone else was seeing this, the witness noticed that everyone was too busy with their assigned jobs. During an UNREP it's "all hands-on deck", so the entire crew has to help out. The witness knew they were the only ships in the area, and it wasn't their helicopter because it was in the hanger, and the tankers helicopter was secured on their flight deck. When he finally decided to tell his shipmates to look at what he was seeing, he pointed at it, and suddenly one light went off, then the next one, then the rest in order, it was as if it realized he seen it. The other shipmates didn't see it.

About a 3 days later the witness had to start his aft-lookout watch, which is standing out on the back of the ship looking for surface and air contacts and report them to combat operations. His watch started right at dusk so there was still light in the sky. The crew was prepping an old stripped out destroyer with a tracking device for bombing exercises the next day which they were going to sink. The destroyer was DIW (dead in the water) engines off so the boat crew could board the old destroyer. The witness remembered that the water was so calm it was like floating on a mirror. One of his friends came down to have a cigarette. The two were so interested in what the boat crew was doing, that he forgot he was on watch. He turned around, and as he did, the very same aircraft he had seen a few nights before was back. There was still light in the sky, enough to realize that the 6 white orbs seemed to be attached to something, but the witness could not see any superstructure. He asked his friend if he was seeing this, and by the look on his face the witness knew his friend was "scared to death". They were both admittedly in shock. The unknown craft was approximately 20ft away, right over the fantail rails not 10 ft over the water. The lights were bright but not blinding, and he guessed, had to be about the size of a large tractor-trailer tire. The craft itself was at least 20ft in length, made absolutely no sound, and was completely still. The craft seemed to have been there for a while before first noticed, and the witness felt that the craft was waiting for him to come out for his watch. These watches are rotated between divisions, and it was his turn to have the night watch. The person who stands the night watch has to be alone all night and into dawn by themselves. No one is allowed on deck at night but the aft lookout. After about 30 seconds of both of the witnesses staring at the craft, it shut it lights off in sequence left to right and just disappeared right before their eyes. The witness did not see the craft fly off in the sky or go underwater, it just vanished. His friend "freaked out" and ran off leaving him out on deck by himself all night.

After the two incidents the witness was terrified and requested to switch to the bridge and take the helm, while his shipmate took his aft-lookout watch. The witness reported the sightings to combat operations after the craft had vanished but was told that nothing appeared on radar and quickly dismissed his report. This case file was from the archives of the Malta UFO Research. The website/group is now defunct, but the case file is archived here for reference.

April 20, 2020, the U.S. Navy Officially Released Three Videos

Three videos were officially released by the U.S. Navy on April 20, 2020, proving the authenticity of UFOs sightings and events. The video footage which was stated as being originally released without official authorization, was taken by Navy pilots, revealing "unexplained aerial phenomena." The three videos released were titled "FLIR.mp4," "GOFAST.wmv," and "GIMBAL.wmv, and were located on its Freedom of Information Act (FOIA) page, which is a repository for documents released under the disclosure agreement of U.S. government as information to be released to the public. The New York Times and To the Stars Academy of Arts & Science, were the first to find and release the videos in 2017 and 2018.

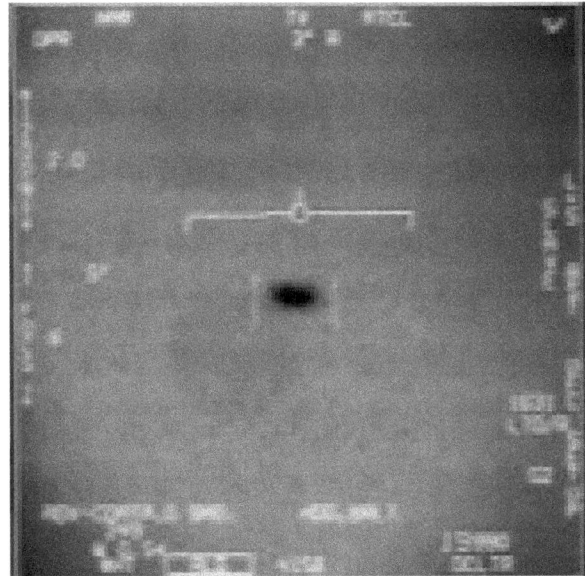

Figure 33 US Navy Stills from Videos
Public Domain

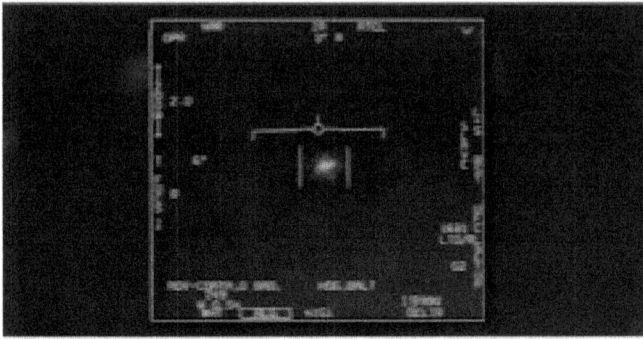

Figure 34 USS Nimitz Photos Public Domain

USS Nimitz, Pacific Ocean

November 2004

Footage Released by the Navy

News broke of the U.S.
Navy's acknowledgment that the 2004 videos of an encounter with a UFO were real with an article published in Popular Mechanics by reporter Tim McMillan.

Two "unknown individuals" told several Naval officers who witnessed the USS Nimitz UFO event, that they were ordered to delete evidence. The report which was published in Popular Mechanics, includes interviews with five Navy veterans who discussed what they had experienced while sailing on the USS Princeton on Nov. 14, 2004, off the coast of southern California. Gary Voorhis who was one of the witnesses to the event, said he was chatting with some of the radar techs on the USS Princeton when he heard them talking about "ghost tracks" and "clutter" on the radar system. This radar is a state-of-the-art Cooperative Engagement Capability system known as (CEC) and AEGIS Combat System.

Gary Voorhis stated that these air control systems were taken down and recalibrated to clear out the false alarms when the tracks got clearer. "Once we finished all the recalibration and brought it back up, the tracks were actually sharper and clearer." These unidentified craft varied in altitude from 60,000 to 80,000 feet, sometimes flying at a mere 30,000 feet at speeds of about 100 knots. Operations Specialist Senior Chief Kevin Day said in the documentary film, The Nimitz Encounters, "Their radar cross sections didn't match any known aircraft; they were 100 percent red. No squawk, no IFF (Identification Friend or Foe)." Day's job was to "man the radars and ID everything that flew in the skies." Day noticed the stranger tracks on the radar on or around Nov. 10, 2004, approximately 100 miles off the San Diego coast. The craft were appearing in groups of 5 to 10 at a time, closely spaced to each other and at an altitude of 28,000 feet. According to Day, the craft were traveling 100 knots tracking south. The Navy eventually sent out fighter jets to get a look at the objects, successfully getting one on video which became the famous black-and-white tape released publicly in 2017.

(Navy, 2015)Atlantic Ocean

Along with that tape, there were two other video recordings recorded by Navy pilots years later and released publicly by the New York Times. The videos known as "FLIR1," "Gimbal" and "GoFast", were originally released to the New York Times and to The Stars Academy of Arts & Science, co-founded by former Blink 182 rocker Tom DeLonge. The first video of the unidentified object was taken on Nov. 14, 2004 and shot by the F-18's gun camera. The second video was taken on Jan. 21, 2015, shows another unidentified craft with pilots commenting on how strange it is. A third video, also taken on Jan. 21, 2015, was released but it is unclear whether the third video was of the same unidentified craft or a third one. After the incident with the "Tic Tac" shaped craft that Gary Voorhis said gave off "a kind of a phosphorus glow" at night while darting around, two "unknown individuals" came aboard the ship and took all of the data recordings. P.J. Hughes, who was miles away from the Princeton stated "They were not on the ship earlier, and I didn't see them come on. I'm not sure how they got there." P.J. Hughes added that he was told by his commanding officer to turn over the recently secured hard drives of the airborne early-warning aircraft, the E-2 Hawkeye. Hughes stated, "We put them in the bags, the commanding officer took them, then he and the two anonymous officers left," Gary Voorhis described a similar situation on the Princeton. "These two guys show up on a helicopter, which wasn't uncommon, but shortly after they arrived, maybe 20 minutes, I was told by my chain of command to turn over all the data recordings for the AEGIS system." This was stated by Gary Voorhis in his interview with Popular Mechanics. It may be convenient to blame "unknown individuals" for the disappearance of the tapes, but Commander David Fravor, one of the pilots who was able to get a close view of the object, said people accidentally erased and recorded over them.

(Filer, 2021)1/25/2021
Former U.S. Navy Commander David Fravor, a former pilot of the MH-53E Sea Dragon, stated that twice while recovering spent practice munitions out of the water, he spotted a weird underwater object. The MH-53E Sea Dragon is the Navy version of the Marine Corps' CH-53E Sea Stallion, which is based at Naval Station Roosevelt Roads, on the island of Puerto Rico. Fravor stated that in the first incident, he saw a "dark mass" underwater as he and his team retrieved a flying practice drone. Fravor described the object as a "big" mass, "kinda circular," and was certain it wasn't a

submarine. Favor's second sighting occurred while retrieving a practice torpedo that was "sucked down" into the depths of the ocean. This occurred in the presence of a similar underwater object as described by him prior. He was never able to find nor retrieve the sought-after torpedo.

Fravor also reveals that a 79-year-old woman contacted him after his sighting went public. The woman told Fravor that her father, a naval officer, was once based at the naval station in San Francisco in the 1950s. She went on to say that when she was a child, her father showed her a telegram that stated "unidentified objects" had been sighted going in and out of the water at a now forgotten set of latitude and longitude coordinates. The woman stated to Fravor that her father had said, "We get these all the time, and it's always in the same area." Fravor stated that the only reason he had seen the now-infamous "Tic Tac" UFO was because it was hovering above a mysterious larger object that was sighted underwater. Fravor described the object as "cross shaped and approximately the size of a Boeing 737 jetliner." The water above it Fravor described as seeming to be "boiling" or "frothing," and stated that the object disappeared after it caught his attention.

Navy Classified Guidelines
In 2019, the Navy issued new classified guidelines on how to report such instances "in response to unknown, advanced aircraft flying into or near Navy strike groups or other sensitive military facilities and formations." The Defense Department also briefed three senators in June, 2019 as part of what appeared to be heightened efforts to inform politicians about naval encounters with unidentified aircraft. Warner's spokesperson indicated that the senators voiced safety concerns surrounding interference that was unexplained that naval

pilots faced, according to Politico, and reported that more briefings were being requested as news surfaced. The Navy needed to revise its procedures for personnel reporting on unusual aircraft sightings.

President Donald Trump said he has been briefed on Navy pilots' reported sightings of unidentified flying objects but remained skeptical of the existence of UFOs. "I want them to think whatever they think. I did have one very brief meeting on it, but people are saying they're seeing UFOs. Do I believe it? Not particularly." This statement was picked up by several news media outlets. In December 2017, Fox News reported that the Pentagon had secretly set up a program to investigate UFOs at the request of former Sen. Harry Reid, who expressed his desire earlier this year for lawmakers to hold public hearings into what the military knows.

December 4, 2020 "Silver Cube"

A leaked Pentagon Photo of a UFO 'Silver Cube' that was hovering over the Atlantic Ocean was released on December 4, 2020. The photo was apparently taken by a military pilot using his cell phone camera off the East Coast of the United States in 2018. The military description of the photo was of an 'unidentified silver 'cube-shaped' object' hovering over the ocean at an altitude of approximately 30,000 to 35,000. The image was reportedly captured by the backseat weapons systems operator of an F/A-18 fighter jet and was speculated as resembling a GPS dropsonde. A GPS dropsonde is an atmospheric profiling device designed to be dropped from an aircraft, which is typically done over a hurricane.

This was ruled out because the object in the photo did not have a GPS transponder dangling from it as a dropsonde would, and dropsondes plunge toward the Earth at 10 to

12 meters per second, they do not hover in mid high-altitude air.

Finally, the 2018 report stated that there was a legitimate possibility that UAP represented 'alien' or 'non-human' technology, which was included in a second, revised report issued by the Unidentified Aerial Phenomena Task Force in 2020. The report stated the possibility that UAP are unidentified craft which are "able to freely move both through the air and underwater, zipping through the ocean undetected and emerging into the air at incredible speeds". The report also contained an "extremely clear" photograph of an unidentifiable triangular aircraft that emerged from the ocean in front of a F/A-18 Hornet fighter pilot. This photo has not yet been released to the public.

Professor Haim Eshed

Press Release: December 5, 2020
Former Head of Israel's Space Program Professor Haim Eshed released the following statement:
"The Aliens Asked Not to Be Revealed, Humanity Not Yet Ready".
In an article released by David Israel Kislev 5781 –
"Prof. Haim Eshed served from 1981 to 2010 as the head of Israel's security space program and over the years received the Israel Security Award three times, twice for confidential technological inventions. In an interview he gave "7 Days", the Shabbat edition of Yedioth Aharonoth, Israel's largest circulation for-pay newspaper, he states, "The aliens have asked not to announce that they are here, humanity is not ready yet." The UFOs have asked not to publish that they are here, humanity is not ready yet. Trump was on the verge of revealing, but the aliens in the Galactic Federation are saying: Wait, let people calm down first. They don't want to start mass hysteria. They want to first make us sane and

understanding. They have been waiting for humanity to evolve and reach a stage where we will generally understand what space and spaceships are. There's an agreement between the US government and the aliens. They signed a contract with us to do experiments here. They, too, are researching and trying to understand the whole fabric of the universe, and they want us as helpers. There's an underground base in the depths of Mars, where their representatives are, and also our American astronauts".

Chapter 9

Ocean Sightings Logged by Ships 1948 Through 1968

UFO sightings were routinely reported and recorded in ship logbooks and not considered a threat to one's job, reputation, or national security. By the 1970s the governments across the world had made the subject of UFOs top secret and out of bounds to the general public. Sightings were still logged but deemed classified. The following are at sea ship reports of anomalies seen and documented by crew members.

(Referrals to meteor, missile or satellite in description is reference to classification or "best description". MERINT (Merchant Ship Intelligence), AFR, ADOIN, ATIC, AFL, AFB refer to sighting report "rule invoked".)

15-Oct-48 22:20 GMT Off Norfolk, VA 36°42'N 74°40'W Atlantic
SS Gulfport/Civil/ 1 object/Nearly as moon/Bright/Straight course/Probably balloon/ 50 min/ 1 witness/Master of ship/SE/ Direct to BB

04-Aug-50 10:00 EDT North Atlantic 39°35'N 72°24.5'W Atlantic
MV Marcala/ Civil/1 object/ovular cylindrical; elliptical 10 ft diameter/Shiny aluminum color and sparkled in the sunlight/Straight course/ 400 to 500 mph/50 to 100 ft above Water; closest to ship was 1000 ft/15 sec or 1.5 min/ 3 witnesses/Master of ship, 3rd Mate and Chief Mate/ 70 deg NE/ Investigated by 2

Intelligence Officers who wrote Information Report

07-Aug-50 08/ 03:50 GMT (19:50 local) From San Francisco to Balboa, Panama Canal 21°50'N 108°45'W/ Pacific/ MV Minerva/ Civil/ Flaming aerial body half size of setting sun glowing frontal dark and backwards/ flaming body/Straight course; unbelievable speed/ Orange-yellow colored tail in the rear/ 2 Meteor Seconds/ 4 witnesses/Master R. Wolport, Chief Mate, A.B. seaman, & Engineer/ 45 deg NE/ Civilian report to Ship Agents in Panama Canal/ Investigated by DIO, 15th Naval District

24-Feb-51 24/ 02:44 GMT 157.06' W 21.32' N 21°32'N 157°6'W Pacific
SS Hawaiian Logger/ Civil /cylindrical/Dark cylindrical object with reddish tail/ Straight course; larger than meteor/ appeared like rocket in controlled flight/ 1 witness/Second Mate Froyland/ ESE /Spot Intelligence Report from Headquarters Pacific Division (MATS) by 14th Naval District ONI, C-3 Restricted Movement Report Center, Hawaiian Sea Frontier

11-Jun-52 09:20 GMT (4:25 local) Returning from the Bahamas 33°0' 75°25' Atlantic
SS Nassau Civil/rectangular/ 20 inches square rectangular yellowish body and a copper green tail/Straight course copper green tail (no exhaust trail)/6 to 10 miles from Ship; 200

yards above sea level Possibly Meteor/ 3-4 seconds/1 Civilian witness/6 to 10 miles from Ship; 200 yards above sea level/Civilian called USAF Inspector General AFR 205-1

14-Sep-52 22:13 local North of Island of Bernholm during Operation Mainbrace in N. Atlantic 55°29'N 14°52.5'E /Danish Destroyer - Operation Mainbrace Danish Navy /3 objects moving in triangular formation; Bluish glowing triangle/Bluish glowing triangle with white light from exhaust/ 2 Objects /changed bearing from 320 to 240 in 5-10 seconds; 932 mph/1,000-meter Altitude/ 1 witness/ Lt. Commander, 2nd in Command of Danish Destroyer 40-50 deg S/ Telegraph from CNO Washington DC

30-Sep-52 Ocean Station UNCLE °' °' Atlantic Ocean Station UNCLE from USCG Coast Guard luminous grayish disc 500 to 600 ft in diameter/ luminous grayish disc/Hovered between the clouds a few minutes then started moving started moving slowly then rapidly to 800 mph/ Altitude = 5000 ft: 2 miles from ship Aircraft few minutes/ 3+ witnesses /Coast Guard Commanding Officer, TANEY (WPG-37) George D. Symon/ 5000 ft Elevation two miles from ship USCG letter to WPAFB

10-Nov-53 21:00 GMT Off Coast of Greenland 60°10'N 37°50'W Atlantic
MS Imanak/ Civil/Light as large as planet in flight as large as Big Planet/Straight course Fast speed/High Altitude/ 4 minutes/ 1 witness/Captain of civilian ship 240 TRUE Danish Naval Headquarters to USAB Greenland ADOIN 1845

04-Apr-54 18:35 GMT 44.10' W 40.38' N 40°38'N 44°10'W Atlantic
MV El Cafetero/ Civil/ 1 Balloon-like Very large/Moving against wind/ White smoke under object/Probably Balloon/2+ witnesses/SGD Hein Pedersen Master of Ship;

observed by several SE Civilian contacted HQ USAF Washington DC who then forwarded to ATIC

12-Aug-54 12/05:55 GMT (20:55 local) Ship anchored off Yoron-Jima 27°2'N 128°27'E Pacific SS Docteur Angier/ Civil/1 object/100 ft diameter/ ellipse/ about length of 5 cent piece held at arm's length; then size of basketball when over ship/ Circle was bright blue, reddish at the inside of the circle and jet black inside the circle of light/Approached ship, slowed down and rose rapidly/no propulsion/200-300 ft above sea/2 witnesses: Mr. Percharde & Kosei Makamoto/ From 0 deg to 30 deg SW/ Civilians were interrogated by USAF Intelligence in Okinawa/rule invoked: AFR 200-2

07-Apr-55 21:15 GMT 65.22'55" W 18.07'5" N 18°7'N 65°22'W Atlantic
CTG 45.2 (Commander Task Group) USN/Circular /larger than moon/bluish silver/ Hovered for 10 seconds, then departed/30 sec 40/ 40 men witnesses/Telegram from CTG 45.2 to COMCARISEAFRON

21-Sep-55 23:00 GMT 36.3' N 73.55' W 36°3'N 73°55'W Atlantic
SS Alcoa Pegasus/Civil/Strange glowing/1 object/Upward slowly and circular motion/ 23 degrees Altitude Possibly Balloon (Moby Dick High Balloon) 10 minutes/23 deg Telegram from ship to CEOMEASTSEAFROM FLYOBRPT / AFL 200-5

21-Jan-56 08:04 GMT 210 miles SE Bermuda 29°40'N 62°0'W Atlantic Danish Ship, Danfjord/ Civil/ round like a star/ 1 Object /was falling, then hovered for 2 seconds, then made abrupt right-angle change to the North/ tail like a red flame /Possibly Meteor/ 8 seconds/ 1 witness/Ship's Second Mate/ 60 deg to 20 deg N /All information was taken from the ship's Log from District Intelligence Officer, 6th Naval

District to Director of Naval Intelligence ONI Instruction 03820.19B & ONI Instruction 08320.17C

20-Sep-56 21/ 05:15 GMT 35 N 131.15' W 35°0'N 131°15'W Pacific
USS Takelma / CTF 31 USN/ 1 rocket shape size of dime/ Black object with no lights/ 600 mph left vapor trail/11 seconds/ 3 witnesses/ CO, OOD, Lookout 190 dgr. / True Telegraph from CTF 31 to RJEDEN/CINCONAD

06-May-57 20:18 GMT 26.38'N 23.22'W 26°38'N 23°22'W Atlantic
SS Hunters Point (Civilian Ship) Civil/ 1 circular shape/ Black center/Straight course/ Brilliant fiery emission forming a ring around it/Possible Comet/ 3+ witnesses/All mates and Radio officer (JB Morton - Master) 8 deg 57 min Altitude 282 dgr True Civilian report - from HYDRT Wash DC to HQ USAF

05-Nov-57 23:40 GMT SSW of New Orleans 27°50'N 91°12'W Atlantic
SS Hampton Roads/ Civil /1 Round glowing object/ edges were brighter than center/Straight course/10 minutes /5 witnesses/Chief Mate, 2nd Mate, 3rd Mate, Radio Operator and Chief Eng. /62 deg NNE From SS Hampton Roads to COMEASTSEAFRON then to USAF HQ MERINT

05-Nov-57 05:10 GMT 200 miles S of New Orleans 25°47'N 89°24'W Atlantic
SS Sebago (USCG) Coast Guard/3 Radar Targets and 1 Egg shaped/ visual Brilliant planet like/ Radar Visual target was flying straight/ 2nd Radar target was Circular pattern/3rd Radar was hovering No vapor trails 2 R1 = 14 to 2 miles / R2 = 22 to 55 miles / R3 = 7 to 175 miles / Visual =3 sec Meteor & anomalous radar propagation R1 = 1 min / R2 = 2 min / R3 = 17 min / Visual =3 sec/ 4 witnesses/Ens. Wayne Schotley/ Navigator Lt. Donald Schaefer/ QM-1 Kenneth Smith/ Radarman

Thomas Kirk/ Visual sighting at 31 deg elev. S and N From COMDR 657th ACWRON to COMDR ADC ENT AFB AFR 200-2

06-Nov-57 20:20 GMT Off coast of N. Carolina 35°22'N 74°14'W Atlantic
Civilian Ship/ Civil/ 1 Large pink illumination/ 1/4 Hemisphere/Like Ionized Gas/Aurora 10 min/ reappeared 4 min later/N from COMCFECR to Air Tech Intel Center

06-Nov-57 01:40 GMT North Atlantic Ocean
Civilian Ship/ Civil/ 1 Bright red /areas 1 /Aurora /North in East West Direction BB Index Card

16-Jan-58 12:00 Local Trindade Island - 600 miles East of Rio de Janeiro °' °' Atlantic
Almirante Saldanha Brazil/Navy/ 1 Disk/Hoax Not Reported/Brazilian Navy/ Available via Photos Available via Photos ONI Information Report based on report from US Naval Attaché at Rio

24-Apr-58 02:45 GMT SE of Newfoundland 45°5'N 49°45'W Atlantic
SS American Hunter/ Civil/ Streak/Mostly white; slightly orange /Straight course/ with a smoke trail/Meteor/From OIN4271 to HQ 26th AIRDIV SE to NW Telex to ATIC from HQ 26th AIRDIV

25-Apr-58 23:33 GMT Atlantic Ocean 38°0'N 54°30'W Atlantic
SS Cocle/ Civil/ Cigar/ Bluish Color/ Straight course/ Very High/ From OFEIN-OI 126 - Message received from Eastern Sea Frontier by Master of ship North Telex from COMCFECR to ATIC

30-May-58 11:45 GMT Approx. 765 miles West of North Africa 24°42'N 32°46'W Atlantic
USCG Ship Coast Guard/Shooting flame & white smoke/from CCGD Seven Miami FL West Telex from USCG to HQ USAF

02-Jul-58 21:15 GMT East of Ascension Island 10°20'S 9°25'W Atlantic
SS African Planet/ Civil/ 1 Bright red missile/ Bright red/Disintegrated in mid-air/No Details/ East Message from ship was sent to Navy and relayed to Intel by Command Post. Report collected by Staff Duty Officer

17-Jul-58 09:10 GMT 76'5 16' 16°0'N 76°5'W Atlantic
USCGC Santa Cecilia Coast Guard/ Surrounded by White haze/ Object heading east, then turned NE/ 3 white lights emanating from main body appeared to be two light rays, one at least 8 degrees in length/8 min/ 4 witnesses/Chief Radio Officer, Second Mate, and 2 ABS on bridge watch varied/ NE /Master sent message to USCG then US Coast Guard COMDT to Bolling AFB

20-Jul-58 03:28 GMT 100-135 miles SE Azores 37°33'N 20°30'W Atlantic Coast
USCG Cutter Eagle Coast Guard/Undetermined Mag/ greater than planet Bright object/ shined through overcast /High speed /Meteor /10 sec/ E USCG /sent report to COM E. AREA

13-Aug-58 23:35 GMT North of Azores 45°20'N 31°0'W Atlantic
SS Marine Progress/ Civil/White pulsating light /Great speed and pulsating every 8 seconds overhead at great height/SSE From COMR ADC to COMDR ATIC/ Report from COMEASTAREA

24-Aug-58 16:28 GMT East of Norfolk, VA 36°42'N 70°25'W Atlantic
M/V Urania/ Civil/ 1 Bright Burning Object/ Bright white light /Falling/ East to NE / 30-45 sec
1 witness/ 2nd Officer Erik Lindquins / NE /Ship reported sighting to USCG New York then to COMEASTAREA

17-Sep-58 8:05 GMT 48'00 35'00 35°0' 48°0' Atlantic
Ocean Station Echo Coast Guard/ 1 Bright white light/ Bright white/Constant position/ angle/2 min /20 deg NE from OSE to COMESTAREA MERINT

09-Oct-58 Not Given Gulf of Campeche, Mexico 19°41'N 93°31'W MX
Shrimp-boat Kaskmer/ Civil/ 1 cylinder with cone 4 ft in diameter/ 4ft submerged and 4ft of cone shaped nose above water/ 8 ft long painted red/ 6 floating objects/Close range/ Floating Buoy/ 1 hour/ 1 witness/Captain of Shrimp Boat Floating Capt. Walter Kirkconnel called USCG vessel by Radio

19-Oct-58 00:40 GMT Atlantic (From Norfolk VA to Bremen, Germany 46°7'N 19°25'W Atlantic MV Coolsingel/ Civil/ 2 spheres attached by tube between thumb and forefinger/ Very white bright sphere with lights like portholes/ Came out of clouds, cross near across the ship, went back into clouds with short reddish tail like a rocket/Quite low/ below clouds (1,000 to 2,000 meters) Meteor 5 to 7 sec/ 2 witnesses/ J. Van Tiel - 2n Officer and Spanish sailor lookout J. Del Rio Fiera Quite low/ below clouds (1,000 to 2,000 meters) Nearly overhead 200 deg MERINT

21-Oct-58 00:45 GMT East of Brazil - Atlantic Ocean 29°2'N 2°28'W Atlantic
M/S Cometa/ Civil /1 Luminous object/ full moon but w/stronger light/ Object flared up for a moment now and then/ Terrific speed/ Probably Meteor/ 5 sec/ 2 witnesses/3rd Mate and other crew/ 10 deg over horizon North/ Telegram received by Embassy CIA/ Commercial Telegraph from Norway
06-Dec-58 21:08 GMT West of Aalesund, Norway/ West of Namsos, Norway 62°30'N 5°4'E Atlantic/ Swedish Fishing Vessels: Vitofors and Kuikkjokk/ Civil /Strong greenish light/Great speed/ Object exploded or

126

disintegrated in flight /Great height/ Meteor/No Details /20 deg over horizon West to SW /US Naval Attached - Oslo Norway

22-Feb-59 05:30 GMT 60 miles East of Puerto Rico 19°10'N 57°30'W Atlantic
MV Colytto/ Civil/ Round 3 ft in diameter Reddish-orange/ 1 Object /was on collision course with ship, made gradual right turn before arriving overhead then gradually turned left and resumed original course/ steaming tail of gaseous flame/ 7,500 ft Missile/ 5 min, 1 min, and 45 sec/ 3 witnesses/Second Mate Mr. Vrie, Fourth Mate Mr. Bakker, Ordinary Seaman (Crew at Dutch merchant vessel) 7,500 ft overhead West from CMDR 5700th AB Group, Albrook AFB Canal Zone to ATIC AFR 200-2A

22-Apr-59 19:50 GMT Between Portugal and Morocco 35°56'N 8°1.5'W Atlantic
SS Ocean Star King/ Fireball/ Very large/ Curve track/25 deg altitude /Meteor/ 4 sec /25 deg altitude North COMEASTAREA NY to ATIC

29-May-59 01:30 GMT Hawaiian Islands - North Pacific 25°12'N 163°50'W Hawaii
USS Newell USN/Flaming meteors/ 1 Object fell ahead of ship 2 From elevation 40 deg to about 10 deg/ Meteor /30 min/From elevation 40 deg to about 10 deg/ From USS Newell direct Telex to ATIC

31-May-59 09:00 GMT Hawaiian Islands - North Pacific 24°0'N 162°50'W Hawaii
USS Plumas County steaming in company of USS Lincoln County and USS St. Clair County USN/White glow first glowed dimly then brightened as if exploding then disappeared/ 2 Meteor/3 sec/3 ships simultaneously E-W USS Plumas County (LST) Telex directly to ATIC WPAFB

30-May-59 19:45 GMT Hawaiian Islands - North Pacific 21°40'N 149°24'W Hawaii

USS Denald County USN/ Bright meteor/ brightened considerable as it fell leaving red trail & finally disintegrating into shower of red pieces/ Meteor 6 sec/ Altitude= 43 deg WNW USS Denald County (LST) Telex directly to ATIC WPAFB

09-Jul-59 01:00 GMT Ocho Rios, Bahamas / SE of Florida 21°13'N 68°24'W Atlantic
SS Esso Stockholm/ Civil/ rocket shape/ Large with lights when passing overhead/ Heavy Tail/ Missile/Altitude= 10 deg 120 deg From Coast Guard Division 7 in Miami, FL

09-Jul-59 01:30 GMT Ocho Rios, Bahamas / SE of Florida 26°5'N 73°5'W Atlantic
SS Caltex Auckland/rocket shape w/ three lights/ 2 Missile/ 10 min/ East from Coast Guard Division 7 in Miami, FL MERINT

09-Jul-59 01:10 GMT Ocho Rios, Bahamas / SE of Florida 26°49'N 73°39'W Atlantic
MS/S Allobrogia/ 1 rocket shape w/ three lights over vessel/ Missile/No Details/ East from Coast Guard Division 7 in Miami, FL MERINT

28-Jul-59 04:13 GMT East of Florida 29°50'N 79°25'W Atlantic
USS Robinson USN/ 1 bright white light/ Object grew larger as it moved across ship; size of a quarter/ white light surrounded by an orange

glow /traveling upward 2 25,000 ft/ Missile /2.5 min/ 3 witnesses/ Student on Quartermaster Watch (Hoskins), OOD, and BM3/ 25,000 ft 250 deg Civilian from Naval Reserve reporting directly to ATIC

09-Sep-59 07:20 GMT 10 miles East of Florida 25°50'N 80°1.5'W Atlantic CGC Androscoggin Coast Guard/ two stage rocket/ Altitude and speed high/ Missile 3.0 min/ USCG SE From USCGC Androscoggin to COMEASTSEAFRON MERINT

30-Sep-59 10/1/ 01:30 GMT West Coast of Florida 28°12'N 85°15'W FL
CGC Nemesis Coast Guard/ Brilliant Flare/2 Flare 30-40 min/ 3+ witnesses/ USCG + numerous other fishing vessels/ 25 deg altitude SE USCG Cutter to USCG Dist. 7th then to CINCONADCOM

06-Oct-59 6:01 GMT Approx. 115 miles East of Cape Canaveral, FL 28°54'N 78°28'W Atlantic
MV Defender/ Civil/ Glowing object surrounded by reddish yellow halo or light/ Flashing white light seen for a few seconds/ Altitude = 28,000 ft/ Missile / USCG East by North USCG to CG Dist. 7th then to CINCONADCOM

21-Oct-59 05:07 GMT 380 miles East of Langley AFB, VA °' °' Atlantic
USS Skywatcher USN/Tear Shape/ Bluish White /Appeared to burn out with vapors following Meteor/ 10 sec /1 witness/Seaman W.J. Lawrence West from COMDR NAYDS McGuire AFB New Jersey to ATIC AFR 200-2

01-Nov-59 23:20 GMT Atlantic Ocean 31°14'N 61°57'W Atlantic
MV Rio de LaPlata/ Civil /1 Bright object White blue of great intensity/ red gas wake remained for 20 min/ Altitude = 22 deg/ Meteor/ 20 min /Altitude = 22 deg/ Falling From ship to CG District 5 then to ATIC MERINT

08-Nov-59 08:25 GMT NE of Hawaii 37°8'N 169°52'W Pacific
SS City of Almaco/ Civil/ 1 Bright object/ Emitted 3 flashes/ Object made low arc and struck water 4 miles North of position/ Probably Meteor/ Falling From ship to Hawaii Air Defense Division MERINT

17-Nov-59 04:40 GMT 320 miles East of Nanteo, NC Not Given Atlantic
USS Skywatcher USN/ 1 Tear drop shape size of pea greenish light/ Faded out with added

brightness/ Smoky trail /Meteor/ 1 to 1.5 sec /2 witnesses/ P.N. Frederick (Standby Lookout) & A.P. Narvesan (Security Watch) South From ship to COMDR NY Air Def Sect McGuire AFB NJ then to ATIC AFR 200-2

25-Nov-59 00:45 GMT Atlantic Ocean 11°55'N 73°15'W Atlantic
SS Santa Monica/ Civil/ 1 Bright fiery object Exploded and burst like roman candle/ Meteor/ 4 sec/ 1 witness/Master of ship/ Elev. = 45 deg East from Ship to COMEASTSEAFRON MERINT

10-Mar-60 02:34 GMT Gulf of Mexico 26°29'N 87°47'W Atlantic
USCG Cutter Cahoone Coast Guard/ 1 Flash bright light with trail/ Probably Meteor / Altitude = 40 deg South/ From CG Cutter to COMDR 2047th AACS SQ Maxwell AFB Alabama

01-Apr-60 09:12 GMT ENE of Hawaii 26°0'N 147°17'W Pacific
SS Lurilane/ Civil /Very bright object/ lighted sea for 12 miles/ lower and slower than meteor/10 sec /1 witness/Ship Descending/ Ship report sent to Hawaii DEF Div. Wheeler AFB MERINT

08-Apr-60 07:30 GMT West of Hawaiian Islands 25°34.5'N 179°53'E Pacific
Commander Task Unit 32.0.4 USN/ 1 Bright object tail was 10 deg long; body appeared to be about the size of a dime/ Blue and red object with blue tail disappeared over clouds Probably Meteor/ Altitude = 25 deg SE to NE from CTU to CINCPACFLT
11-May-60 07:06 GMT WSW of Hawaiian Islands Pacific
USS Navasota USN/ 1 Meteor like object tail same size as head/ pea size red head and bluish white tail /Diminished in brilliance until it disappeared /Probably Meteor/ 5 sec /No

Details South from NAVASOTA to PACAF
OPNAVINST 3820.9 / CINCPACFLTIN ST 3820.3

25-Jun-60 00:34 GMT Vicinity of Ascension
Island
Operations Range Vessel - ORV Whiskey USN/
1 Bright glow Light was white or yellowish and
appeared to radiate in 360 deg circle 100 yards
from Data Cassette to be recovered in the
ocean/ Flare 10/ sec 5 /1 Aircrafts with 2
witnesses and one ship with 3 witnesses: PAA
Diver, RCA Photographer & Recovery Specialist
Floating 100 yards from boat and 25 deg off
starboard Manager of Operations Planning
sent letter to AF Missile Test Center, Patrick
AFB, Florida

25-Jun-60 00:15 GMT Vicinity of West Indies
18°0'N 54°10'W Atlantic
ESSO Manchester/ Civil /1 Fairly brilliant
/Brighter than Mars/ Yellow/ Climb at fast
speed 2 Missile - Titan Missile from Cape
Canaveral /4 MERINT Reports from Sea / 1
from Airlines /1
from Nassau Cay/ 1 San Juan CG 120 deg Ship
to COMEASTAREA MERINT

25-Jun-60 00:12 GMT Vicinity of West Indies
22°6'N 69°1'W Atlantic
FPZT Junon/ Civil/ 1 UFO /Missile /Not
Reported /No Details /W to E Ship to
COMEASTAREA MERINT

25-Jun-60 00:15 GMT Vicinity of West Indies
29°23'N 65°35'W Atlantic
Polyclipper Leap/ Civil/ 1 Bright body/ Missile/
10 min /110 deg Ship to COMEASTAREA
MERINT

25-Jun-60 00:17 GMT Vicinity of West Indies
26°5'N 69°28'W Atlantic
SS Soya Pacific/ Civil/ 1 Satellite/ Missile/ 1
witness no Details /Ship to COMEASTAREA
MERINT

05-Jul-60 16:30 GMT 145'50E 39'20N = Aircraft
Location 23°59'N 175°20'E Pacific
NWA Civilian A/C and Navy Ship USS Noble
USN/ 1 Steady white light size and shape of
faint star/ High rate of speed/ Missile/3
witnesses/ Pilot from Northwest Airlines and
US Navy Ship (Quartermaster and Seaman) Alt
= 30 deg NW to SE Ship report to CINCPACFLT

29-Jul-60 19:04 GMT Atlantic Ocean Location
Not Given Atlantic
SS Exilona/ Civil/ 1 Oblong silver gray /Contrail
lasted 3 hrs./ Missile/ 3 min/ 1 witness/
Master of ship/ ship to COMASDEFORLANT

06-Aug-60 23:00 GMT Atlantic off Florida
29°57'N 79°42'W Atlantic
USCG Ship Bethex Coast Guard/ 1 White
colored flare with green projectile 20 ft from
flare Stationary Projectile being 3 inches in
diameter and 3 ft long/ 2 to 3 miles NE of flare
Sonobuoy/ 2 hours/ 1 witness/from Coast
Guard Seven to COMDR 32 NORAD RGN DIR of
l Intel Dobins AFB GA MERINT

10-Aug-60 22:54 GMT Atlantic Ocean 24°10'N
49°17'W Atlantic
NRI Itacielo/ Civil/ 1 Green light /Moving Fast
Followed by white smoke/ Missile /11 min /
1 witness no Details/ Alt = 41 deg NW to SE
from ship to COMANDER East Area MERINT

26-Aug-60 19:40 GMT North of Midway Island
33°55'N 137°32'E Pacific
USS Higbee USN/ 1 Object w/brightness of 1st
mag star/ 2 Echo I/ 10 min/ 1 witness no
Details Alt = 10 deg SW to NE from USS Higbee
to COMSEVENTHFLT/ 5th AF Japan / From
COMBARPAC to ATIC CIRVIS
29-Aug-60 11:16 GMT Vicinity of Taiwan
24°41.5'N 119°38.8'E FA
USS Boyd USN/ 1 Round/ size of pinhead/
white object; bright as planet/ Disappeared
over horizon /Echo I /4 min/ 1 witness no
Details/ Alt = 35 deg NE from Ship to US

Taiwan JOC Taipei Taiwan CINCPACFLT INST 3820.3

03-Sep-60 11:24 GMT Vicinity of Taiwan 24°20'N 114°0'E FA
USS Boyd USN/ 1 Round size of pinhead white object; bright as planet/ Disappeared in low clouds/ Echo I /11 min /3 witnesses/ Capt. B.U. Smith, CDR H.C. Behner, and Watch Personnel ESE CINCPACFLT INST 3820.3

08-Sep-60 08:15 GMT North of Hawaiian Islands 30°0'N 162°39.5'W Pacific
USS Newell USN/ 1 Brilliant Object/ Yellow green / Traveling S to N 2 /Probably Meteor /5 sec /1 witness no Details/ N From ship to COMBARPAC MERINT

16-Sep-60 10:10 GMT Close to Puerto Rico 11°15'N 78°53'W PR
SS Santa Ana/ Civil/ 1 UFO/ Lost from sight in low clouds /1960 Epsilon Reentry/ Not Reported /1 witness/ Ship Alt = 3.2 deg from 20 deg to 80 deg From Ship to COMASDEFORLANT MERINT

16-Sep-60 10:00 GMT Close to Puerto Rico 19°20'N 75°3'W PR
USS Geiger USN/ 1 Disk Shape/ Large Brightly illuminated/ 200 mph/ appeared hovering at times and changing direction /1960 Epsilon Reentry/ 30 min/ 1 witness/ Ship Medium Altitude E from Ship to COMASDEFORLANT MERINT

24-Sep-60 00:55 GMT Off North Coast of Labrador (2nd ship at 57'20"N 60'30"W) 58°35'N 62°40'W Atlantic SS Gannet & SS Algareen/ Civil/ 1 Cylinder-like Larger than Aircraft with 5 port holes like lights along-side / Object approached the surface of the ocean at an angle, hit, and disappeared under the water. No Exhaust / 15 miles south of ship/ Probably Meteor/ 2 witnesses/ 3rd Mate on the Gannet and First Mate on the Algareen/ Alt

= 45 deg ends a 0 deg E to W Canadian Ships reporting to COMDR Goose NORAD Sector Gosse AFB Labrador Canadian

18-Oct-60 09:26 GMT NE of Bermuda 37°14.8'N 59°3.5'W Atlantic
USS Mullinnix (Part of Task Group 20.9) USN/ 1 Bright Light same size and intensity as Cappella Stationary for 5 min then moved to NE/ 10 min /Only 1 witness amongst group of ships /NE ONI Report by LCDR USN H.C. Qudley Not Reported

05-Dec-60 11:48 GMT Japan 34°7'N 140°27'E Pacific
USS Gunston Hall USN/ 1 Spherical object size of pea/ bright red / Fell straight down; appeared to breakup and extinguish rapidly/ tail streaming behind diminishing in brightness, length about width of silver dollar / Meteor/ 4-5 sec / 2 witnesses/ A.H. Clark Lt. USN on Duty as officer of the Deck and V.A. Walker GMC USN Junior Officer of the Deck/ Elev. = 50 deg N /Ship report to 5th Air Force Fuchu Air Station Japan; Naval Message to PACAF Hickam AFB/CINCPACFLT/CNO Moon Dust

06-Dec-60 11:18 GMT Japan 33°50'N 142°20'E Pacific
USS Gunston Hall USN/ 1 Bright object/ Pinhead size Incandescent white with white tail Disappeared by breaking up and extinguishing/ Tail length about width of baseball /Probably Meteor/ 2 sec/ 2 witnesses/ J.S. Eckenrod LTJG USN on Duty as Officer of the Deck and by 4 deck seamen on watch Elev. = 30 deg WSW Naval message from ship to CG USAFPAC/Others
19-Dec-60 09:21 GMT SE Guam 11°41'N 147°4'E Pacific
USS Haverfield USN/ 1 Satellite like/ 1st Magnitude star/ Straight Flight/ Satellite /19 min/

3 witnesses/ CO, XO, OMD-PSOS/ Elev. = 15 deg SE/ Ship to Andersen AFB Guam then ATIC based on AFR 200-2 CINCPACFLT INST 3820.3

10-Jan-61 17:44 GMT Atlantic 19°48'N 73°40'W Atlantic USS Franklin D. Roosevelt USN/ Speed = 3960 knots /Picked by Radar SPS 12/ Track Missile 1 min 1 / South CINCLANTFLT INST 03360.2C

14-Jan-61 05:17 GMT Pacific Ocean 53°35'N 162°0'W Pacific SS Ocean Mail/ Civil/ 1 Bright object /Great brilliance/ Moving Easterly/ Meteor/ 1 witness/ Master/ Elev. = 55 deg East /Ship report to COMNAVDEFEASTPAC MERINT

24-Jan-61 22:05 GMT 3 Ships in Atlantic - US East Coast 23°55'N 58°25'W Atlantic SS Zephyros/ Civil/ 1 Illuminated body Surrounded by cloud of smoke/ Falling / Missile/3 min /1 witness/ Master /Elev. = 60 deg SE MERINT

24-Jan-61 22:05 GMT 3 Ships in Atlantic - US East Coast 24°15'N 56°25'W Atlantic German Tanker Arie Boettger/ Civil/ 1 Irregular flying object/ Glowing and smoking/ Missile / 5 min/ 1 witness/ Master/ NW to SE /German Tanker Ship to USCG in New York then to COMEASTAREA MERINT

24-Jan-61 22:08 GMT 3 Ships in Atlantic - US East Coast 29°14'N 49°57'W Atlantic SS Andalien/ Civil /1 Strange, brilliant object like moon with big blue-white luminescence/ Running speedy in sky 265 deg to 230 deg /Missile/ 5 min/ 1 witness/ Master /Elev. = 20 deg Ship to USCG in New York then to COMEASTAREA MERINT

29-Jan-61 09:12 GMT Atlantic 25°43'N 48°22'W Atlantic USS Pecos USN/ 1 Apparent satellite/ Same magnitude as Vega/ Traveling very fast/ Very High/1 witness/ No Details/ 130 deg True Ship to COMASDEFORLANT NORVA

MERINT

06-Feb-61 11:00 GMT Philippine Sea 20°0'N 134°0'E Pacific SS Master Daskalos/ Civil /1 Bright object/ Huge/ like shooting star /Falling in EW direction breaking into 2 pieces/ Burning up in atmosphere / Meteor / 1 witness/ Radio Officer/ East to West /Ship to COMWESTSEAFRON

08-Feb-61 19:57 GMT 149'42 23'34 23°34'N 149°42'W Pacific USS John S. McCain USN/ 1 Bright object/ White with appearance of 1st mag star/ 1 Object was seen again 2 hour and 8 min later / Satellite/ 8 min/ 1 witness no Details/ Elev. = 30 deg NE/ Ship to CINCPACFLT

11-Feb-61 05:00 GMT Pacific Ocean 9°36'N 145°15'W Pacific SS Lone Star/ Civil/ 1 Round white object/ moving slowly/ 1 witness/ No Details /Elev. = 50 deg 45 deg /Ship radio telephone to COMNAVDEFEASTPAC MERINT

21-Feb-61 01:05 GMT 150 nautical miles East of Cape Canaveral 28°15'N 77°41'W Atlantic SS Erihio/ Civil /1 Object/ was bright and circular /1 ft in diameter/ Amber colored/ object rising to sky very rapidly; followed by 2 wings forming/ parabola in center of which was main body/Probable Missile/ 90 sec /6witnesses/ Master plus 5 crew members/ SW to NE/ Ship report to Coast Guard Division Seven MERINT

28-Feb-61 22:25 GMT Atlantic Ocean 52°40'N 34°55'W Atlantic USCG Cutter Barataria Coast Guard /1 Bright object/ Brightness of 1st mag star/ Straight course High altitude/ Satellite/ 8 min /1 witness/ No Details/ Elev. = 10 deg SE/ From USCG Cutter to OS Charlie then to COMEASTAREA MERINT

05-Mar-61 00:10 GMT Atlantic Ocean 42°55'N 43°30'W Atlantic

MSTS American Angler/ Civil/ 1 Bright object/ Brightness of 1st mag star /1 Orbiting object/ From Ship to CG Radio Station Argentina then to COMEASTAREA

05-Mar-61 19:00 GMT Atlantic Ocean 49°38'N 20°26'W Atlantic M/V Prince Willem Van Oranj/ Civil/ 2 Missile like/ One going about 90 deg the other 100 deg Streaks of smoke still visible 1 hr. later/ 40 min/ From ship to Coast Guard NY then to COMEASTAREA MERINT

15-Mar-61 19:42 GMT Atlantic Ocean 37°35'N 9°14'W Atlantic USS Card USN/ 1 Bright object/ Magnitude 1.7 / Straight course / High altitude / 8 min 1 witness/ No Details /Elev. = 27 deg/ Bearing 263 deg Merint Report to CINCLANTFLT MERINT

03-Apr-61 02:30 GMT Pacific Ocean 34°34'N 132°15'W Pacific
SS President Grant/ Civil /1 Elongated object/ High rate of speed/NE MERINT

13-Apr-61 19:55 GMT Far East 28°51'N 131°45'E Pacific
USS Coral Sea USN/ 1 Round object; bright light Brighter than most stars /moving bright light, very fast/ No trail, or exhaust/ 8 min /Elev. = 10 deg W to E /Naval Message to 5th AF CINCPACFLT 3820.3
26-Apr-61 11:40 GMT East Indies 4°20'N 111°3'E Atlantic
USS Eversole USN/ 1 White object 2nd Mag star/ Straight course/ Satellite/ 12 min
4 witnesses/ CO, OOD, JOOD, and Lookout/ Elev. = 15 deg E /Naval Message from ship to CNO & USAF Pac Hickam AFB

09-May-61 18:45 GMT Far East 22°22'N 124°0'E Pacific
USS James E. Kyes USN/ 1 Bright round object size of pin head/ White/Straight course; Speed approx. 10 deg arc per min/ 2 witnesses/ OOD and JOOD /Elev. = 33 deg 120 deg /Ture Naval

Message from ship to COMNAVFORJAP CINCPACFLT 3820.3

13-May-61 11:50 GMT East Indies 4°14.5'N 106°18.5'E Atlantic
USS Saint Paul USN/ 1 Round size of second mag star/ White /Traveled at high rate of speed across sky /1.5 min/ 5 witnesses/ LCOL R.E. Dunlap USAF, LCDR GA Murphy USN, LT R. Karger USNR, LT G.E. King USN, LT NHS Walczak USN/ Elev. = 60 deg then to 15 deg/ Naval Message from ship to Clark AFB CINCPACFLT INST 3820.3

14-May-61 08:00 GMT Atlantic Ocean 14°10'N 95°10'W Atlantic
SS Penmar/ Civil/ 1 Bright object/ Bright as 1st mag star/ Object with controlled flight too high to note shape or size; stopped twice, then made easy turn and headed S/ Had neither rocket nor meteor shape/ 8 min/ Elev. = 40 deg N then S/ Report from ship to COMWESTAREA MERINT

16-May-61 00:30 GMT Atlantic Ocean 56°17'N 50°52'W Atlantic
Civilian Ship - United States Lines/ Civil/ 1 Bright Object/ Straight Course; high speed/ Possible Satellite/ Ship radio to Ocean Station Bravo. OSB telex to COMEASTAREA then to CINCLANTFLT MERINT
16-May-61 05:31 GMT Atlantic Ocean 43°46'N 37°0'W Atlantic
Navy CINCLANT Fleet USN/ 2 two white lights/ White blinking light / Speed est. Mach 2/ 80,000 ft/ 217 deg Telex/not reported

01-Jun-61 08:45 GMT Far East 34°2'N 164°17'E Pacific
Ocean Station Victor Coast Guard /1 Star like object/ Brighter than star/ 2 High altitude Echo I Satellite/ 8 min/ Elev. = 70 deg then 19 deg ESE /120 deg True Ship report picked up by CG District Fourteen - Naval Message from OSV to COMWESTAREA MERINT

02-Jun-61 11:00 GMT Off San Clemente Island, Wilsons Cove °' °' CA
MV Morning Star/ Civil/ 1 Round to Oblong/ Changing back and forth/ Bright Incandescent bluish/ tinge/ Speed faster than star but slower than A/C/ Disappeared into cloud/ 15 min East to West Merchant Vessel sent via Coast Guard then to COMNAVDEFEASTPAC MERINT

10-Jun-61 11:25 GMT 131'27 17'53 17°53.5'N 131°27'E Pacific
USS Rowan USN/ 1 Bright object/ Brilliance of 1st Mag star/ Moving at orbital speed
2 Echo I /Satellite/ 8 min/ Elev. = 60 deg to 25 deg SE Naval Message from ship to COMSEVENTHFLT Not Reported

10-Jun-61 11:18 GMT Far East 19°25'N 116°20'E Pacific
SS Korean Bear/ Civil/ 1 Bright object Brilliant as Jupiter/ Emitted fire and fragments while continuing on course/ 8 min/ Elev. = 60 deg to 20 deg SE /Report from ship to Naval Defense Forces Eastern Pacific Not Reported

06-Jul-61 14:35 GMT Pacific Ocean 22°40'N 154°14'W Pacific
SS Gisna/ Civil /1 Brilliant object /White, same brilliance as star Vega/ Probably Echo I Satellite/8 min/ Elev. = 22 deg to 12 deg ENE/ M/V Ship report to Coast Guard District 14 then to Hawaii Air Defense Division Wheeler AFB Not Reported

11-Jul-61 01:50 GMT Atlantic Ocean 30°30'N 76°25'W Atlantic
SS Pueblo/ Civil /1 Diamond shape with light at each point/ red and 2 white/ Very high/ speed 2 Medium altitude Aircraft/1 witness/ Master /East Ship report to Coast Guard District 5 then to CINCLANTFLT MERINT

13-Jul-61 00:40 GMT Atlantic Ocean 23°16'N 69°27'W Atlantic

SS Olympic Breeze/ Civil/ 1 Fireball/ Object exploded with a big bang then extinguished itself and a circular white cloud was spreading blown by wind Particles fell into sea/ Meteor/ Merint Report from ship to COMGANTS MERINT

18-Jul-61 14:00 GMT Pacific Ocean 0°49'S 170°24'W Pacific
SS M. E. Lombardi/ Civil/ Steady white light/ Straight course /motion 2 /Satellite/ 20 min Elev. = 20 deg then to 15 deg SSE/ Merint Report from ship to COMWESTSEAFRON MERINT

24-Jul-61 07:00 GMT Pacific Ocean 24°50'N 152°20'W Pacific
SS Flying A California/ Civil /1 Bright object white; brilliance of 1st mag star/ Straight course motion 2/ Satellite - Echo I 8 min/ 1 Elev. = 50 deg then 60 deg NNE Merint Report to COMHAWSEAFRON MERINT

31-Jul-61 07:21 GMT Pacific Ocean 28°13'N 121°16'W Pacific
USS Lansing USN/ 1 Round Size of pinhead White object; bright as 1st mag star/ Straight course passed over the ship /2 Echo I Satellite/ 8 min/ Elev. = 90 deg to 0 deg ENE/ Ship report to Hawaii Air Def Division Kunia Annex

04-Aug-61 08:50 GMT Pacific Ocean 42°30'N 173°8'W Pacific
USS Forster USN/ 1 Round Size of pea White object; bright as 1st mag star/ Straight course passed overhead /Probably Satellite/ 15 min/ 8 witnesses/ V.L. McBeth SA USN Lookout, OOD, CO, and 5 bridge watch standers/ Elev. = 8 deg to 90 deg NE /Naval Message from ship to CINCPACAF CINCPACFLT INST 3820.3

06-Aug-61 08:00 GMT Far East 22°27'N 175°19'E Pacific
Civilian Ship - United States Lines/ Civil/ 2 Bright objects Light Pulsating/ 2 objects; one

behind the other pulsating; speed = 9 times of echo 2/ 10 sec/ 3 witnesses/ Civilian and 2 AB's while on watch/ Elev. = 53 deg North to South/ Not Reported

18-Aug-61 16:00 GMT Pacific Ocean 24°53'N 150°22'W Pacific
SS Lurline/ Civil/ 1 Vapor trail N/A Moved southerly over vessel and spread rapidly/ Vapor Trail from horizon to horizon/ Low altitude vapor trail/ Probably Missile/ 5 min/ South Merint Report to Hawaii Air DEF Division Kunia MERINT

04-Sep-61 07:34 GMT Pacific Ocean 34°47'N 166°19'W Pacific
NAVAIR/SAR Ship/ Civil/ 1 Star like /Size and brightness of 1st Mag star/ White/ Satellite 8 min/ Elev. = 10 deg East/ Merint Report to Hawaii Air DEF Division Kunia MERINT

13-Sep-61 10:00 GMT East Coast of US 32°0'N 68°54'W Atlantic
USS Sarsfield USN/ 1 White light similar to planet/ Resembled small disc/ Expanded rapidly then exploded forming thin, orange-colored mass 3-time diameter of sun/ 1 light approached rapidly from North above and behind low cumulus cloud bank/ Rocket Launch 5-7 sec/ 4 witnesses/ Navigator, OOD, JOOD, and Lookouts/ S Naval Message from ship to CNO/CINCLANTFLT

13-Sep-61 09:39 GMT East Coast of US 36°25.5'N 71°30'W Atlantic
USS Randolph USN/ 1 Small elongated orange colored light; then round silverish/ light appeared 6 to 8 times diameter of moon/ Light became large and brighter resembling hazy moon. It blossomed out in a burst of bright orange light globular in shape/ 1 Rocket Launch 5-7 sec/ 2 witnesses/ A.E. Clarke QM1 USN, JOOD LTJG T.M. Atkins /Elev. = 45 deg to 70 deg 320 deg true to 40 deg/ Naval Message from ship to CNO/CINCLANTFLT Not Reported

23-Oct-61 14:30 GMT Pacific Ocean 58°14'N 143°55'W Pacific
MV Chena Civil/ 1 White flash/ No Tail/ 2 witnesses/ Captain and Pilot N Naval /Message from CCGD 17 to COMALSEAFRON AFR 200- and OPNAV INST 3820.9/ CGD17 Operations Instruction 29- 61

23-Oct-61 15:09 GMT Pacific Ocean 58°46'N 144°29'W Pacific
USS Harris County USN/ 1 Bluish white flash/ Straight course/ Meteor type/ tail or trail of smoke/ Meteor /3 sec/ 2 witnesses/ Lookout & Officer on watch/ 65 deg /Naval Message from CCGD 17th to COMALSEAFRON AFR 200- and OPNAV INST 3820.9/ CGD17 Operations Instruction 29- 61

30-Oct-61 03:59 GMT Pacific Ocean 30°30'N 158°35'E Pacific
USS Noble USN/ 1 Object w/brightness of 2nd mag star/ Straight course, observed to make slight changed of course to the right/ Satellite Echo/ 8 min/ Elev. = 66 deg SE/ Naval Message from USS Noble to COM HAW SEAFRON Not Reported

03-Nov-61 04:58 GMT Pacific Ocean 24°15'N 170°15'W Pacific
USS Durant USN /1 Round object/ Pin-head size/ White/ 2 Echo I satellite/ 8 min/ Elev. = 70 deg NE/ Naval Message from USS Durant to COMHAWSEAFRON Not Reported

09-Nov-61 06:30 GMT Far East 30°5'N 175°0'E Pacific
USC GC Winnebago Coast Guard/ 1 Star-like object/ Brightness of 2nd mag star/ Straight course 2 Echo I satellite /15 min/ NE/ Merint report from CGC Winnebago to CRMEDWESRQ UCH to COMHAWSEAFRON MERINT

12-Nov-61 04:31 GMT Pacific Ocean 28°34'N 163°42'W Pacific

USS Durant USN /1 Round object Pin-head size /White/Straight course/ 2 Echo I satellite 9 min/ 3 witnesses/ LCDR USN, LTJG USNR OOD, QMI USN ASST NAV /NE/ Merint report from USS Durante to Haw AIR DEF DIV Kunia Annex Hawaii MERINT

15-Nov-61 22:32 GMT Atlantic Ocean 28°4'N 69°5'W Atlantic
SS Venture/ Civil/ 1 Cone shaped object/ Extreme brilliance; sparks emitted from below object Tail executing slow parabolic curve with comet tail/ Missile shape/ - Transit 4B /4 min/ 1 witness/Master of ship/ Elev. = 40 deg then to 20 deg SE/ Merint report from civilian ship to CCGD 7 (Coast Guard District 7) MERINT

18-Nov-61 05:24 GMT Pacific Ocean 40°12.5'N 173°45'W Pacific
USS Durant USN /1 Round object size of pinhead/ White object/ blinking as A/C warning beacon/ Straight course/ Aircraft shape/ 3 min/ 3 witnesses/BM3 USN Watch stander, LTJG AJN OOD, LCDR USN CO /Stop at Elev. = 30 deg N /Telex to COMHAWSEAFRON/ Not Reported

05-Dec-61 05:14 GMT Pacific Ocean 40°32'N 166°59'W Pacific
USS Wilhoite USN/ 1 Round object /Size of planet/ Straight course/ Probable Satellite Echo I/5 min/ 8 witnesses/HSN USN Lookout, CO, XO, OOD and 4 others/ Elev. = 20 deg S /Naval message from USS Wilhoite to USAF PAC/CINCPACFLT/Other; then MERINT report sent by HAW AIR DEF DIV KUNIA Annex to Hickam AFB CINCPACFLT INST 3820.3/ MERINT

15-Dec-61 00:40 GMT Atlantic Ocean 50°31'N 17°30'W Atlantic
USS General W.O. Darby USN /1 Missile like Blue-white-red-green/ Sparking/ observed when object crossed bow; maintained glide path/Not Reported/ No Details/ NW/ Merint

report from USNS GEN WO Darby to COMASWFORLANT MERINT

15-Dec-61 18:25 GMT Pacific Ocean 40°0'N 174°0'W Pacific
USS Lansing USN /1 Bright object like satellite /Object with 2nd mag star brightness/Straight course/ 5 min /1 witness/Elev. = 75 deg then 80 deg Merint report from USS Lansing to Air Div. Kunia Facility Hawaii MERINT

20-Dec-61 18:25 GMT Pacific Ocean 40°7.5'N 173°5'W Pacific
USS Durant USN/ 1 round object size and mag of Venus/Straight course/Probable Satellite Echo I /5 min/ 4 witnesses/CO, Navigator, QM1, and Asst. Navigator /Elev. = 62 deg then to 30 deg NE /Naval message from USS Durant to Hickam AFB Hawaii/ COMALSEAFRON/ Others; then MERINT report from HAW AIR DEF DIV KUNIA Annex Hawaii to USAF MERINT

14-Jan-62 18:51 GMT Pacific Ocean 31°31'N 171°52'E Pacific US CGC Winnebago Coast Guard/ 1 Satellite like/Altitude very high/ Echo Satellite/Not Reported/ 1 witness/NE/ Merint/ Report from USC GC Winnebago to COMHAWSEAFRON MERINT

15-Jan-62 11:15 GMT Atlantic Ocean 38°30'N 74°20'W Atlantic
SS President Van Buren/ Civil/ 9 Blue-green objects/ Size of 1st and 2nd Mag Stars; varying in brilliance and had red halos/ 5 Flying together in various random patterns/Objects flew on various courses, NNE then E then SE then S/ Medium to high altitude /Probably Aircraft/Not Reported/ 1 witness /No Details/ NNE then E then SE then South/ Ship notified CG District 5/ Telex from CCGD Five to CINCLANTFLT/ USAF Not Reported

29-Jan-62 23:40 GMT From South Carolina to Puerto Rico 32°6'N 67°5'W Atlantic
SS Federal Monarch/ Civil/ 1 Globular object 1/4 diameter of moon resembling cluster of flashing lights/ 4 with trail of luminous haze 2 Missile like/5 min /1 witness/No Details/ Elev. = 25 deg SE /SS Federal Monarch reports to USCG NY who reports to COMEASTAREA

29-Jan-62 23:38 GMT Submarine Base in Puerto Rico Puerto Rico Atlantic
CGC Sagebrush Coast Guard /1 Cone shaped loom; L shaped lights; appeared triangular in shape Brightness of 1st Mag star/ had 2 lights flashing and a halo type cloud encircled/ the object Est. 50,000 ft/ Missile test/ AFMTC Test 119/1 5 min 12 No Details Elev. = 30 deg E CGC Sagebrush located at sub base tender in PR to COMGANTS Not Reported

02-Feb-62 21:45 GMT Atlantic Ocean 48°12'N 28°26'W Atlantic
SS American Shipper/ Civil/ 1 Bright object/Straight course/90 arc in 6-7 sec/ With a long and bright tail visible through clouds/Meteor like/ 6-7 sec/2 witnesses/Captain WA Woodman and 2nd Officer HW Schonn/ Elev. = 20 deg then to 5 deg SE /From Ship to Ocean Station Delta. Naval Message from OSD to Govt. Hydro Office Not Reported

13-Feb-62 07:32 GMT Pacific Ocean 29°10'N 163°20'W Pacific/
USS Lansing USN/ 1 Satellite - Echo I/ Brightness of 1st Mag Star/ High Elevation/ Echo satellite Not Reported/SE /Telex & Naval Message from Air Div. Kunia Facility Hawaii to ATIC

21-Feb-62 09:10 GMT Pacific Ocean 21°16'N 121°40'E Pacific
USS Vernon County USN/ 2 Spherical Pea size White/ Small object above and right of large object/ 2 Satellite - Echo I /5 min/ 4

witnesses/Longton (CO), Psassantino (XO), Yates (OOD), and Hammond (QM2)/ Elev. = 22 deg then 05 deg E and SE/Naval message from USS Vernon to CG Fifth AF PACAF CINCPACFLT INST 3820.3

22-Feb-62 10:23 GMT Pacific Ocean 21°16'N 121°40'E Pacific
USS Noble USN/ 1 Bright object/ Brightness of 1st Mag Star/Straight course/ Echo I Satellite 5 min/ 4+ witnesses/Navigator, CO, OOD, JOOW and 1,000+ men/ Elev. = 35 deg then 20 deg SE/ Naval message from USS Noble to USAF PAC / 5th AF FUCHU AS Japan Not Reported

26-Feb-62 01:05 GMT Pacific Ocean 9°14'N 90°32'W Pacific
SS Hawaiian Retailer Kroz/ Civil/ 1 Bright object/ Straight course/ Echo I Satellite/ 3 min/ 1witness/ Dobbs 2nd Officer/ Elev. = 59 deg then 34 deg SE/ Merint report from ship to COMWESTSEAFRON Navy San Francisco MERINT

05-Mar-62 08:25 GMT Pacific Ocean 58°28'S 174°4'E Pacific
USS Arneb USN/ 1 Bright object/ Brightness of 1st Mag Star/Straight course motion/ Echo I Satellite/Elev. = 20 deg E /Naval Message from USS Arneb to COMNAVSUPPFOR Antarctica Not Reported

17-Mar-62 13:34 GMT NE of Hawaiian Islands 32°48'N 118°45'W Pacific
USS Estes USN/ 1 Ball of flame/ Size of dime/ Light blue/Straight course with tail 5 times length/ Meteor like/ 2 sec/ 3+ witnesses/CO, EO and others/ Elev. = 30 deg to 25 deg E/ Naval message from USS Estes to CG Vandenburg AFB CINCPACFLT INST 3820.3

03-Apr-62 07:00 GMT Atlantic Ocean 40°55'N 46°48'W Atlantic

USS Lt. Geo W.G. Boyce USN/ 1 Bright object/ Brightness greater than 1st Mag Star/Straight course & high/ speed 1,000 MPH /Extremely high altitude/ Satellite/ 6 min/ 1 witness/No Details/ E /Merint Report from Ship to EASTSEAFRON MERINT

07-Apr-62 17:30 GMT Pacific Ocean 22°40'N 125°0'E Pacific
USS Noble USN/ 1 Bright object/ Brightness of 2nd Mag star/Straight course/ Satellite/ 5 min/ 3+ witness/OOD, JOOD, and bridge watch personnel/ Elev. = 10 deg then 5 deg NNE/ Naval Message from USS Noble to Hickam AFB/ Not Reported

12-Apr-62 00:45 GMT Atlantic Ocean 31°11'N 65°43'W Atlantic
USS Maloy USN/ 1 Bright object /White/Straight course/ speed = 1,000 MPH/ Meteor/ Not Reported/ 1witness/ No Details/ 310 deg/ Naval message from USS Maloy to CINCLANTFLT

12-Apr-62 15:31 GMT Pacific Ocean 10°52'N 149°59'W Pacific
USS Newell USN/ 1 Small star/ Small pea/ White /Straight course/ Satellite/ 5 min/ 3 witnesses/ Morin (XO), Southwick (OOD), Simmons (SM2)/ Elev. = 28 deg to 20 deg 150 deg/ Naval message from USS Newell to USAF Pack Hickam AFB CINCPACFLT INST 3820.3

19-Apr-62 11:24 GMT Pacific Ocean 14°59'N 149°2'E Pacific
USS Thomaston USN /1 Round object/ Pea size /White/Straight/ circular course/ No Tail/ Satellite/ 5 min/ 1 witness/No Details/ Elev. = 45 deg to 25 deg SE/Naval Message from ship to CNO & CINCPACFLT Not Reported

22-Apr-62 07:50 GMT Near Wake Island 18°45'N 166°0'E Pacific
USS Thomaston USN/ 1 Round object/ Pea size/ White/Straight circular course/Fast speed/No Tail/ Echo I/ 5 min/ 5 witnesses/OOD, JOOD, OPS Officer, Lookout and Quartermaster/
Elev. = 35 deg then 10 deg NE/ Naval message from ship to CINCPACFLT Not Reported

22-Apr-62 16:16 GMT Near Wake Island 19°6'N 168°12'E Pacific
USS Thomaston USN /1 Round object/ Pea size/ White/Straight circular course/Short reddish tail/ Echo I/ 5 min/ 4 witnesses/OOD, JOOD, Lookout and Quartermaster/ Elev. = 48 deg to 4 deg SE/ Naval Message from ship to CNO/CINCLANTFLT Not Reported

23-Apr-62 07:15 GMT Near Midway 27°9'N 176°34'E Pacific
COMPHIBRON One USN/ 1 Star like/ White/Straight course/ Echo I /5 min/ 2 witnesses/Rohan (Navigator) and Burdick (QM2)/ Elev. = 40 deg to 30 deg E/ Naval message from COMPHIBRON ONE to CG Fifth AF CINCPACFLT INST 3820.3

26-Apr-62 08:04 GMT Pacific Ocean 30°0'N 138°30'W Pacific
A/C and Ocean Station November Coast Guard /1 Flare like/ Green/Straight course/Meteor/ 5 sec/ 1 witness/No Details/ E Telex from Air Div. Kunia Facility Hawaii to PACAF Hickam AFB CIRVIS

02-May-62 04:20 GMT 137'19 28'31 28°31'N 137°19'W Pacific
Ship Report from Hawaii Base USN/ 1 Falling Star like object/ Brightness of 1st Mag star/ White Straight course/ Small tail/Meteor/ 5 min/ 3 witnesses/ 2 Navigators and Quartermaster/Elev. = 20 deg from 18 deg to 30 deg/ Telex from Air Div. Kunia Facility Hawaii to PACAF Hickam AFB Not Reported

02-May-62 05:25 GMT Pacific Ocean 28°23'N 137°30'W Pacific
COMPHIBRON One USN/ 1 Bright object/ Larger than 1st Mag Star/ Red, white, and blue/ Straight course/5 min/ 1witness/ Furgerson (Staff Watch)/ Elev. = 10 deg to 14 deg S/ Naval message from COMPHIBRON ONE to WADF Hamilton CINCPACFLT INST 3820.3

11-May-62 02:00 GMT Pacific Ocean 42°50'N 155°20'E Pacific
SS Washington Bear/ Civil/ 2, 4 engine aircraft with red markings/ Mid wing/ swept tail/ Red V on tail and red star/ 6 Flew low over ship/100 ft above surface then pull up to steep climb/Aircraft/ Not Reported/ 1 witness/No Details/ Merint report from ship to Ocean Station Victor/ Naval Message from OSV to COMWESTAREA CCGD MERINT

14-May-62 07:10 GMT Pacific Ocean 5°21'N 157°28'W Pacific
USS Forster USN/ 1 Round Solid with glow/ Size of 1st Mag Star/ Orange white with yellowish cast/ Straight course/ No Tail/Satellite/ 5 min/ 3+ witnesses/OOD, Boatswain of watch, Lookout, and entire bridge watch/ Elev. = 50 deg then to 52 deg S to SE/ Naval message from ship to USAF PAC CINCPACFLT INST 3820.3

17-May-62 08:35 GMT Pacific Ocean 30°30'N 174°17'E Pacific
CGC Winnebago Coast Guard/ 1 Satellite looking/ White/Straight course/2,000 MPH/Echo I/15 min/130 deg/ Merint Report from Ship to CINCPACFLT MERINT

22-May-62 06:45 GMT Pacific Ocean 25°29'N 172°52'W Pacific
USS Stone County USN/ 1 Bright shiny object/Straight course/ Satellite/ Not Reported/

Elev. = 25 deg then 0 deg SE /Merint report from Ship to Air Div. Kunia Facility Hawaii MERINT

13-Jun-62 17:00 GMT Pacific Ocean 8°8'N 105°5'E Pacific
USS Mathews USN/ Bright object; pear shaped/ Brighter than 1st Mag Star/2 to 3 times the size of any planet/ Brighter than Venus/white cloudy glow appeared to grow around center/ 1 Object disappeared and reappeared; gaining altitude rapidly; object appeared to be superimposed upon a course of 4 or 5 small lights/Formalhaut Star over 5 hours/ Elev. = 0 deg then 73 deg/ Naval message from USS Mathews to CINCPACFLT Not Reported

04-Jul-62 09:08 GMT Pacific Ocean 11°45'N 174°52'W Pacific
USS Forster USN/ 1 Round object/ 1st Mag Star/ Solid glow; bluish/ white/ Straight course/ No Tail/Satellite/ 5 min /3 witnesses/OOD, CO, and Bridge Watch /Elev. = 45 deg then 05 deg NE/ Naval Message from Ship to CINCPACFLT HICKAM CINCPACFLT INST 3820.3

06-Jul-62 05:30 GMT Pacific Ocean FM 005/ 061 Pacific
USS Falgout USN/ 1 Round; Star like object/ Bright as 1st Mag star/ White/ Straight course Satellite/ 5 min/ 5 witnesses/ Lookouts, OOD, CO and Navigator/ Elev. = 50 deg to 0 deg NE MERINT Report CINCPACFLT INST 3820.3

12-Jul-62 10:50 GMT Pacific Ocean 33°22'N 165°44'W Pacific
USS Wilhoite USN/ 1 Round object/ 2nd Mag star/ White/Straight course/Satellite – Echo/ 5 min/ 4 witnesses/OOD, Lookouts, and CO/ Elev. = 38 deg to 14 deg East/MERINT Report from Ship to COMWESTSEAFRON MERINT

12-Jul-62 12:55 GMT Pacific Ocean 32°52.2'N 165°23'W Pacific
USS Wilhoite USN/ 1 Round object/ 2nd Mag star/ White/ Straight course/ Satellite - Echo 5 min/ 3 witnesses/OOD and Lookouts/ Elev. = 75 deg to 11 deg East/ MERINT Report from Ship to COMWESTSEAFRON MERINT 1

13-Jul-62 05:40 GMT Pacific Ocean 18°25'N 55°47'W Pacific
USS Falgout USN/ 1 Round/ Brightness of 1st Mag Star; pinhead/ White/ Straight course/ Satellite – Echo/ 5 min/ 4 witnesses/OOD, Lookouts, and CO/ Elev. = 38 deg to 5 deg ENE Naval Message from ship to CINCPACAF Hickam CINCPACFLT INST 3820.3

14-Jul-62 02:15 GMT South Pacific - East of the Marques Islands 6°9'S 110°38'W Pacific
SS Southern Eagle/ Civil /1 Satellite looking Luminous object / Straight course/ Satellite/ 1 min/ 1 witness/Master of Ship/ S to N /Civilian Merint Report MERINT

18-Jul-62 14:00 GMT Pacific Ocean 0°49'S 170°24'W Pacific
SS ME Lombardi/ Civil/ 1 Satellite looking/ White/ Straight course/ Satellite/ 5 min/Elev. = 20 deg then 15 deg NNW to SSE/ Civilian Merint Report to COMWESTSEAFRON MERINT

29-Jul-62 10:47 GMT Pacific Ocean 34°42'N 163°42'E Pacific
Ocean Station Victor Coast Guard/ 1 Bright object/ Brighter than 1st Mag Star/ Blue white/ Straight course/ 5 min/Elev. = 80 deg S OSV Report to Air Div. Kunia Facility Hawaii Not Reported

29-Jul-62 21:18 GMT Atlantic Ocean 31°6'N 32°18'E Atlantic
MS Ferngate/ Civil/ 1 Bright star like or satellite/ Straight course/Satellite/ 5 min /1 witness/ First Officer Nygard/ Down to Horizon

East Civilian ship to Navy Hydrographic Office then to NASA then to BB/ Not Reported
31-Jul-62 09:10 GMT Pacific Ocean 34°0'N 163°50'E Pacific
CGC Matagorda Coast Guard/ 1 Bright object/ Brighter than 1st Mag Star/ White/ Straight course/ Satellite/ 5 min/ Elev. = 50 deg ESE /From USCG ship to COFS USAD and COMWESTAREA Not Reported

08-Aug-62 08:35 GMT Pacific Ocean 33°52'N 163°56'E Pacific
Navy Ship - Name not reported USN/ 1 Bright object/ Brighter than 1st Mag Star/ White/ Straight course /Satellite/ Elev. = 79 deg SE /Telex from NIFV to HBDR/ Not Reported

22-Aug-62 15:35 GMT Pacific Ocean 4°12'S 154°4'W Pacific
USS Durant USN/ 1 Bright object; star like/ 1st Mag Star/ Straight course/ Satellite/ 5 min/ 2 witnesses/Ike (Exec) and Harris (OOD)/ Elev. = 43 deg to 25 deg NE/ Naval Message from USS DURANT to Hickam AFB CINCPACFLT INST 3820.3

23-Aug-62 05:05 GMT Pacific Ocean 7°18'S 153°41'W Pacific
USS Durant USN/ 1 Bright object/ Brightness of 2nd Mag star/ Straight course / Satellite/ 5 min/ 2 witnesses/ Davidson (BM of Watch) and Lookout/ Elev. = 30 deg to 25 deg NE/ Naval Message from USS DURANT to Hickam AFB CINCPACFLT INST 3820.3

23-Aug-62 06:11 GMT Pacific Ocean 7°34'S 153°39'W Pacific
USS Durant USN/ 1 Bright object/t Brightness of 2nd Mag star/ Straight course/ Satellite/ 5 min 3 witnesses/Laipply (LTJG), IKE (OOD), Mclennan (JOOD)/ Elev. = 20 deg to 56 deg S Telex from Air Kunia Facility to ATIC/ Not Reported

29-Aug-62 17:30 GMT Pacific Ocean 35°4'N 144°10'E Pacific
CGC Matagorda Coast Guard/ 1 Bright object/ Less than 1st Mag Star/ Straight course/ Satellite 5 min/ Elev. = 10 deg N/ Telex from CGC Matagorda to COMWESTAREA and COFS USAF then from Kunia Air Div. to ATIC MERINT

30-Aug-62 08:20 GMT Pacific Ocean 23°34'S 163°44'W Pacific
USS Durant USN/ 1 Bright object/ Size of Venus / Straight course but speed seemed irregular/5 min/ 2 witnesses/ Simpson (OOD) and Dale (Lookout)/ Elev. = 5 deg to 10 deg East/ Naval Message from USS Durant to Hickam AFB CINCPACFLT INST 3820.3

01-Sep-62 21:30 GMT Atlantic Ocean 3°8'S 35°56'W Atlantic
SS Hyperion - Civilian Greek Tanker/ Civil/ 1 Rocket/ White star like color/ Very Fast, straight like motion/ Rocket or Missile like/ 2 min/ NE /Ship report to Coast Guard Station in San Juan then to CG District 7 then to COMCARIBSEAFRON Not Reported

02-Sep-62 18:19 GMT Pacific Ocean 34°35'N 163°3'E Pacific
Ocean Station Victor Coast Guard/ 1 Bright object/ Bright as 1st Mag Star/ Straight course/ Elev. = 38 deg NE/ Telex from OS Victor to COFS USAF, COMWESTAREA, others Not Reported

10-Sep-62 20:52 GMT South China Sea 22°29'N 118°8'E Pacific
SS Illinois /Civil/ 1 Satellite junk burning/ Passed in front of ship and disappeared into clouds/ Sparks trailing object/ Meteor/ 4-6 sec / North Marine Data Report sent to Naval Oceanographic Office by ship, then forwarded to NASA, then ATIC Marine Data Report to HO

16-Sep-62 06:20 GMT Pacific Ocean 37°20'N 124°8'W Pacific

SS Californian/ Civil/ 1 Meteor like; bright object/ Yellowish green/ flare/ Straight course/ Meteor/ 3 sec/ West/ MERINT Report by ship MERINT

16-Sep-62 18:35 GMT Pacific Ocean 33°55'N 163°48'E Pacific
CGC USS Matagorda Coast Guard/ 1 Bright spherical object /Bright as 1st Mag star/ Straight course/ East/ Telex from CGC Matagorda to COMWESTAREA and COFS USAF Not Reported

05-Oct-62 05:08 GMT Pacific Ocean 11°5'N 165°19'W Pacific
USS Lansing/ USN/ 1 Bright object/ Star like/ Great speed; straight course/ Satellite /Not Reported / No Details/ Elev. = 60 deg to 45 deg NE/ From ship to CTU 8.3.6 MERINT - JANAP 146D

27-Oct-62 23:15 GMT 50.5 miles East of Cape Canaveral 28°27.5'N 79°34.5'W Atlantic
M.V. Boxhill/ Civil/ 1 Ball shaped orange light / Ascending Eastward /Tail of light had 3 colors: red, white and blue / Missile like/ 3+ witnesses/ Chief Officer, Captain, and other officers/
Elev. = 2 deg then 40 deg East/ Marine Data Report sent to the US Navy Hydrographic Office then to ATIC Marine Data Report to HO

02-Nov-62 02:50 GMT Atlantic Ocean 43°25'N 67°0'W Atlantic
T.M.S. Irvin Glenn/ Civil/ 1 Bright object/ Half size of moon to size of orange/ Greenish/ Straight downward/ Meteor/ 5 sec /1 witness/ Second Officer/ Elev. = 45 deg to 3 deg/ Descending Letter from Second Officer to US Gov. forwarded to ATIC/ Not Reported

17-Nov-62 09:20 GMT Pacific Ocean 11°59'N 160°49.5'W Pacific
MV Aconis/ Civil /1 Bright object/ Red / Breaking up and falling/ Elev. = 20 deg Falling/

Civilian Merint Report got to ATIC via Air Div. Kunia Facility MERINT

20-Nov-62 10:00 GMT Pacific Ocean 23°46'N 69°19'W Pacific
US CAGS Hydrographer/ Civil /1 Brilliant object/ Brightness of Jupiter / Straight course/ Satellite 6 min/ 1 witness/ COS/ NE/Letter from US Dept. of Commerce Coast and Geodetic Survey ship Hydrographer to the US Navy Hydrographic Office then to NASA then to ATIC US Hydrographic Office

23-Nov-62 20:23 GMT Pacific Ocean 16°8'N 131°58'W Pacific
USS Tulare USN /1 Round Pea sized/ White/ Straight course/ high speed/ 5 min/ Elev. = 35 deg then 20 deg SE/ Naval Message from USS Tulare to Clark AFB/ Not Reported

03-Dec-62 18:30 GMT 29'45 64'45 64°45'N 29°45'W Atlantic
Navy Ship USN/ 1 Burning object/ Disintegrating; straight course/ Trailing fire/ Meteor/ West Naval Message from COMBARFORLANT to USAF HQ/ Not Reported

04-Dec-62 03:45 GMT Gulf of Mexico 25°5'N 85°4'W Atlantic
SS Esso Lima/ Civil/ 1 Rocket like/ Straight course/ Vapor trail shooting upward/ Elev. = 15 deg then 30 deg/ South Merchant Vessel Merint Report went to US CG District 8 which then send Naval Message to CINCNORAD, COMEASTSEAFRON, & Others CIRVIS

31-Dec-62 05:05 GMT Pacific Ocean 2°18'N 162°9'W Pacific
CGC USS Matagorda Coast Guard/ 1 Bright object/ Bright as 2nd Mag Star/ Straight course/ Satellite/ Elev. = 39 deg 55 deg/ Merint report from USS Matagorda to Air DIV Kunia Facility then to ATIC MERINT

19-Jan-63 13:51 GMT Pacific Ocean 43°59'N 124°51'W Pacific
Oregon State University Oceanographic Ship/ Civil/ Round bright light/ small particle black particle followed by a ball of fire/ Fell into water/ Followed by a ball of fire/ Meteor/ 2 sec/1 witness/ Captain Ben Gertulla/ SW Captain reported to Portland Air Defense Sector via Dept. of Oceanography - OSU, then to ATIC/ Not Reported

31-Jan-63 06:00 GMT Atlantic Ocean 33°30'N 56°55'W Atlantic
SS Dunkirk and 3 Air Visuals/ Civil/ 1 Bright flash/ White/ Moving in southern sky toward horizon then explosion/ light up sea for 5- 10 sec/ Short tail of sparks or debris / Meteor 5-10 sec/ 4 witnesses/ 3 Air Visuals and one Merchant Vessel/ reported via Ocean Station Echo Elev. = 40 deg to 0 deg/ Exploding /Telex from OS Echo to COMEASTAREA MERINT

07-Feb-63 12:30 GMT Pacific Ocean 15°34'N 100° 8'W Pacific
USS Marysville USN /1 Satellite / Orbital motion/ Satellite/ 5 min/ Elev. = 42 deg NE/ Letter from CO of USS Marysville to US Naval Oceanographic Office then to NASA then to ATIC US Oceanographic Office

16-Mar-63 07:05 GMT Pacific Ocean 3°45'S 168°19'W Pacific
USS Durant USN/ 1 Blinking light/ Pea size/ White/ Straight course/ blinking period of 35 and 20 sec/ 5 min/ 1 witness/ Lookout/ Elev. = 0 deg S/ Naval Message from USS Durant to Hickam AFB then to COMHAWSEAFRON CINCPACFLT INST 3820.3

19-Mar-63 01:40 GMT Atlantic Ocean 17°55'N 57°40'W PR
SS Plainsman Civil/ 1 Object/ exploded in sky /Object fell into water and exploded again/ Fell into water/ MERINT report from ship to US

Debbie Ziegelmeyer

Coast Guard in PR then to COMCARIBSEAFRON then ATIC MERINT

28-Mar-63 02:24 GMT Pacific Ocean 32°17'N 117°22'W Pacific
USS Sculpin USN/ 1 Circular/ Head of Pin/ Bright as star; no A/C lights/ Straight course / Satellite - Echo I /5 min/ 3 witnesses/ Lt. Parks, Lt. JG Howse, Lt. Akers/ Elev. = 53 deg to 30 deg East Naval Message from USS Sculpin to March AFB, Calif.; Telex from CINCPACFLT to ATIC CINCPACFLT INST 3820.3

28-Mar-63 21:50 GMT Atlantic Ocean 35°0'N 48°0'W Atlantic
USCG USS McCullogh Coast Guard/ 1 Star like/ Fast moving/ star/ Satellite/ Elev. = 62 deg to 15 deg 60 deg/ Naval Message from USCG Cutter to CINCLANTFLT/ Not Reported

16-Apr-63 06:30 GMT Pacific Ocean 24°43'N 166°23'W Pacific
USS Cossatot USN/ 1 Star like object/ Glowing object/ Straight course / very fast/ 20,000 ft Elev. = 20 deg East /Telex from COMHAWSEAFRON to ATIC MERINT

18-May-63 07:57 GMT Pacific Ocean 46°15'N 174°15'W Pacific
USS Savage USN / 2 Bright object; like satellite/ 2nd Mag star/ Like falling star; same object appeared 1.5 hours later on same course/ Meteor like/ 10 sec / 80 deg/ Telex from NOBARPAC to ATIC MERINT - JANAP 146D

21-May-63 20:10 GMT Atlantic Ocean 40°2'N 10°17'E Atlantic
USS Gearing USN/ 1 Rocket /1 Object was climbing produced red vapor trail /Missile like 2 min Letter from USS Gearing to COMSIXTHFLT/ Not Reported

26-May-63 09:00 GMT Gulf of Mexico 29°14'N 93°25'W Atlantic Tug Sandra/ Civil/ 1 Bright object/ Bluish white light/ Erratic course; very

high rate of speed/NE Ship reported to Coast Guard then Naval Message from CG District 8th to CINCNORAD, COMEASTSEAFRON, others MERINT

30-May-63 07:41 GMT Hawaiian Islands 30°50'N 169°0'W Pacific
USS Lansing USN/ 1 Bright object/ 2nd Mag star/ 1 Object disappeared and reappeared twice/ High altitude/ 5 min / Elev. = 45 deg then 12 deg NW/ Merint report sent from ship to COMBARPAC. Telex reports to ATIC from COMPARPAC and AIR DIV Kunia Facility Hawaii MERINT JANAP 146D

13-Jun-63 16:54 GMT Pacific Ocean 34°0'N 164°0'E Pacific
USS Chautauqua Coast Guard /1 Round bright object/ White / Ascending /with red trace/ low on horizon /Ascending/ Merint report from ship to Ocean Station Victor and from OSV to CINCPACFLT MERINT

14-Jun-63 23:01 GMT Ocean Station Echo Position Not Given Atlantic
Ocean Station Echo Coast Guard/ 1 Satellite/ 1st Mag Star / Moving NE / Satellite - Echo I/ Elev. = 75 deg NE/ Merint report from OSE to COMEASTAREA MERINT
187 48

16-Jun-63 11:20 GMT Pacific Ocean 39°9'N 152°6'E Pacific
SS Tetela/ Civil /1 Bright object/ Irregular flashing light/ Did not appear to be on straight course/ Probably Aircraft/E /Merint Report from SS Tetela to USCGC Chautauqua to CINCPACFLT MERINT

17-Jun-63 06:31 GMT Pacific Ocean 28°30'N 162°20'W Pacific
USS Lansing USN/ 1 Bright object/ Similar to 1st Mag Star/ Similar to Planet Venus / Straight course / Satellite - Echo 1/ 5 min/ Elev. = 85 deg to 5 deg NE/ Merint Report from ship to

COMBARPAC & from Air Div. Kunia Fac. Hawaii
MERINT JANAP-146D

17-Jun-63 08:30 GMT Pacific Ocean 28°0'N
162°0'W Pacific
USS Lansing USN/ 1 Bright object/ Similar to
1st Mag Star/ Straight course /Satellite - Echo
1/   5 min/ Elev. = 20 deg to 35 deg NE/ Merint
Report from ship to COMBARPAC then to Air
Div. Kunia Fac. Hawaii HADOC-D 1074 MERINT
JANAP-146D

17-Jun-63 16:38 GMT Pacific Ocean 36°25'N
155°52'E Pacific
USS Halsey Powell USN /1 Spherical Baseball at
arms-length /White object with intense light/
Traveling at high speed/ No Tail/ 3 witnesses/
Fisher (OOD), Stewart (BM3), & Varlamos (SN)/
Elev. = 85 deg to 0 deg SE/ Naval Message
report from ship to CINCPACFLT OPNAV INST
3820.9

20-Jun-63 09:30 GMT Between Japan and CA
39°18'N 175°28'W Pacific
MS Ascanius/ Civil/ 7 Bright objects like
satellite/ Straight course/ seen seven times/ 2
Echo I satellite/ SE/ Ship Visit Report collected
by Nautical Information Branch of USN
Oceanographic Office sent to NASA then ATIC
US Oceanographic Office Ship Visit Report

21-Jun-63 00:42 GMT Gulf of Mexico 15°56'N
81°5'W Atlantic
MV Pensacola Civil/ 1 Bright object as bright as
Polaris in flight /Straight course /Satellite -
Echo I/ 5 min / Elev. = 16 deg to 5 deg NE/
Marine Data Report from civilian ship collected
by NIB and sent to NASA US Oceanographic
Office/ Marine Data Report

21-Jun-63 08:30 GMT Pacific Ocean 33°12'N
168°33'E Pacific
USCGC Matagorda Coast Guard/ 1 Bright
object/ bright as 1st mag star / Straight
course/

Elev. = 47 deg 340 deg true/ Merint Report
from CG ship to USAFHQ/Others MERINT

04-Jul-63 11:25 GMT Far East 26°4'N 127°53'E
Pacific
USS Cook USN/ 1 Bright object/ 1st mag star;
pinpoint/ Straight course / Possible 1961 Beta
Theta Satellite/ 5 min/ 4 witnesses/ Griffin
(CO), Morandini (OOD), Gaddy (Lookout),
Hines (VJOOD)/ Elev. = 10 deg to 17 deg NE/
Naval message from USS Cook to
COMNAVPHIL CINCPACFLT INST 3820.3

04-Jul-63 13:10 GMT Far East 25°2'N 127°15'E
Pacific
USS Cook USN/ 1 Bright object /1st mag star
moving / Straight course /Satellite - Echo I
5 min/ 5 witnesses/ Griffin (CO), Morandini
(OOD), White (XO), Hines (VJOOD), Caudel
(BM1)/ Elev. = 10 deg to 54 deg SE/ Naval
message from USS Cook to 5th AF FCHPHAS
Japan CINCPACFLT INST 3820.3

15-Jul-63 12:06 GMT Far East 22°2'N 117°0'E
Pacific
USS Marshall USN/ 1 Bright object/ 1st mag
star moving/ White / Straight course/ Satellite
- Echo I /5 min/ 2 witnesses/ Hopper (CO),
Butler (OOD)/ Elev. = 60 deg to 20 deg SE/
Naval
message from ship to CLARK AFB CINCPACFLT
INST 3820.3

21-Jul-63 09:10 GMT Pacific Ocean 15°34'N
115°8'W Pacific
MS Yugala/ Civil/ 1 Ball like/ Glaring white;
conical reflection seen on sea surface/ Straight
course/ Conical tail with lower part white,
upper part flaming red / 10 miles/ Meteor/ 3
sec 1 witness/ 2nd officer/ Elev. = 20 deg to 10
deg 260 deg T/ Merint report from ship to
COMEASTAREA and then to COMWESTAREA
MERINT

25-Aug-63 07:15 GMT Atlantic Ocean 35°02'N 48°15'W Atlantic
Ocean Station Echo Coast Guard/ 1 Satellite/ Satellite like/ 120 deg Telex from OS ECHO to COMEASTAREA Not Reported

25-Aug-63 08:00 GMT Pacific Ocean 31°57'N 174°25'E Pacific
USCGC Matagorda Coast Guard/ 1 Bright light/ 2nd Mag star/ flashing / Course change from 130 deg to 150 deg./ 5 min/ Elev. = 40 deg to 50 deg 130 deg then to 150 deg /Telex from USCG Cutter Matagorda to COFS USAF/ CG Div. 14 & others MERINT

21-Sep-63 03:43 GMT Pacific Ocean - Off Coast of Oregon 45°22'N 135°34'W Pacific
Picket Vessel PV-3 from Portland Air Defense Sector and Civilian Airliner USN/ 1 Brightly glowing round object/ 1st Mat Star/ Red / Passed overhead; straight course/ Satellite - Echo I/ 5 min/ Elev. = 30 deg East/ Telex from POADS ADAIR AFS Oregon to ATIC AFR 200-2

23-Sep-63 10:25 GMT Pacific Ocean 18°02'N 161°3'E Pacific
USS Deliver USN/ 1 Bright streak/ Basketball/ Brilliance lit up ship and horizon/ Blue with red sparks and long streak/ Descended in a straight course/ Meteor/ 1-2 sec/ 4 witnesses/ OOD, BM, & two Lookouts/ Elev. = 30 deg to 0 deg/ Descending/ Naval Message from ship to CINCPACFLT Not Reported 2

25-Sep-63 05:25 GMT Pacific Ocean 35°05'N 152°6'W Pacific
SS Sunny Lady/ Civil/ 1 Bright light/ Straight course/ Satellite/ 10 min/ East/ Radio Communication from ship to COMNAVDEFEAST PAC then to ATIC MERINT

25-Sep-63 11:06 GMT Pacific Ocean 34°14'N 125°29'W Pacific
USS Lowe USN/ 1 Round object/ 1st Mag star/ Red/ Straight course with luminous vapor

swirling about object; Vapor is 20X size of object/ Missile like/ 4 min/ 4 witnesses/ LTJR RJ Sands - USN OOD, LCDR RJ Edris - USN CO, ENS JE Benton USN JOOD, ER BM2 - USN BMOW/ Elev. = 0 deg to 25 deg West/ Naval Message from USS Lowe to COM $ AF HAM AFB California CINCPACFLT INST 3820.3

09-Oct-63 03:30 GMT Pacific Ocean 32°07'N 117°17'W Pacific
USS Calvert USN/ 1 Bright light /Magnitude of Jupiter/ Straight course/ Satellite/ 5 min/ 2 witnesses/ Kennedy (Navigator) and Austing (SN)/ Elev. = 15 deg East/ Ship letter to USN Hydrographic Office, then HO Nautical Information Branch to NASA then to ATIC US Oceanographic Office Publication 606

04-Nov-63 05:46 GMT Pacific Ocean 40°00'N 174°0'W Pacific
SS Falgout/ Civil /1 Bright light /Star like or planet like / 60 deg arc in 45 seconds/ 5 min/ Elev. = 15 deg NE/ Civilian Merint report to Air Div. Kunia Facility HAW then to ATIC MERINT

11-Nov-63 05:53 GMT 173' 40' 40°00'N 173°40'W Pacific
SS Falgout/Civil /1 Bright light/ Star like or planet like / 45 deg arc in 55 seconds/Elev. = 25 deg 185 deg/MERINT Report to COMBARFORPAC then Kunia to ATIC MERINT JANAP 146-D 2

02-Dec-63 17:50 GMT Farallon Islands Pacific Not Given CA
USS Firedrake USN /1 Missile like object/ Object made turn from SW and headed in South direction/ Trailing plume similar to Regulus type missile/ 50 miles from object; elevation was 10,000 ft to 30,000 ft/ SW to S/ Naval Message from USS Firedrake to COMCENSECTESTSEAFRON/ MERINT evaluation from COMNAVDEFEAST PAC MERINT

10-Dec-63 05:40 GMT Atlantic Ocean 40°20'N 17°40'W Atlantic
USS Savage USN/ 1 Spherical/ pulsating white/ Straight course/ 2 witnesses/ OOD and Lookout East Telex from Air Div. Kunia to ATIC with Merint report JANAP 146-D MERINT

26-Jan-64 19:11 GMT Pacific Ocean 35°58'N 164°15'E Pacific
USS Matagorda Coast Guard /1 Bright light/ 1st Mag Star/ White/ Straight course / Satellite/ Elev. = 30 deg 170 deg/ Telex from CGC Matagorda to USAF Headquarters/CCGD 14/others MERINT

05-Feb-64 00:09 GMT Atlantic Ocean 65°00'N 28°43'W Atlantic
Navy Commander Task Group USN/ 1 Bright light/ Straight course/ Satellite/ 20 deg Telex from COMBAREFORANT to CNO/CINCLANTFLT/others/ Not Reported

17-May-64 06:30 GMT Atlantic Ocean 38°02'N 42°38'W Atlantic
SS President Jackson/ Civil/ 1 Bright object/ Straight course/ flying Medium altitude/ SE/ Telex from COMASWFORLANT to CINCNORAD MERINT

18-May-64 02:00 GMT Far East 34°00'N 164°00'E Pacific
Ocean Station Victor - Radar Sighting Coast Guard/ 1 Radar detection only/ Radar Change courses; speeds of 150 knots/ 29 to 70 miles/ Aircraft 39 min / Radar Detection Case /280-330 deg/ Telex from OSV to CINCPACFLT/ CINCNORAD/ COFS USAF/ Others MERINT

18-May-64 04:45 GMT Pacific Ocean 30°00'N 140°00'W Pacific
Ocean Station Victor Coast Guard/ 1 Satellite/ Straight course/ Satellite/ 5 min /
Elev. = 40 deg to 50 deg NE /Telex from OSV to COMWESTAREA MERINT

06-Jun-64 22:00 GMT 600 miles NW of Ascension Not Given Atlantic
SS Norma C. Penn/ Civil/ 1 Bright light/ Bright as brightest star; similar to Star Rigel/ Appeared to flash; white light/ Erratic motion; object in flight crossing own path; Stood still; suddenly speed up and move away; flash or flicker; disappear and reappear/ Star Vega like/ 8 min/ 3 witnesses/ Master, 2nd Officer and 3rd Officer/ Elev. = 60 deg Northerly; 110 deg/ Captain reported sighting to US Consul in South Africa who then send letter to US Air Attaché then to ATIC forms

25-Jul-64 08:05 GMT Pacific Ocean 41°27'N 164°16'W Pacific
SS Dona Maru/ Civil/ 1 Flare rocket /Blue / Rising and falling from horizon/ Flare/ Rising and falling from Horizon 130 deg/ Merint report to Air Division Kunia Facility then Telex to ATIC MERINT

26-Jul-64 02:58 GMT Pacific Ocean 15°01'S 132°31'W Pacific
Greek Steamer SS Frixos/ Civil/ 3 Twin satellites/ Bright as planet Jupiter/ One was red and green; the other was red and white; third was white/ Orbital motion; distance between 2 objects was steady; after 2 min another satellite joined/ Aircraft & Satellite/ 5 min /
1 witness/Chief Officer/ Elev. = 19 deg 2-West 1-NE/ Merint report to COMWESTAREA then to Air Division Kunia Facility then Telex to ATIC MERINT

05-Aug-64 04:43 GMT Pacific Ocean 30°00'N 140°00'W Pacific
US CGC Pontchartrain Coast Guard 1 Bright light/ Satellite like/ Straight course /Elev. = 66 deg then 32 deg 20 deg/ Merint report from USCG Cutter to CINCNORAD then to ATIC MERINT

11-Aug-64 05:01 GMT 168'00 35'10 35°10'N 168°00'W Pacific
USS Newell USN/ 1 Bright object/ 1st Mag Star/ Satellite like/ Straight course/ Elev. = 30 deg 60 deg/ Telex from COMBARFORPAC to COMHAWSEAFRON MERINT JANAP 146-D

25-Oct-64 01:30 GMT Atlantic Ocean 66°00'N 27°40'W Atlantic
Navy Ground Visual / Ship Unknown USN/ 1 Satellite like/ Straight course/ 270 deg/ Telex from ADMINO COMBARFORLANT to CNO/CINCLANTFLT/ CINCNORAD/ Not Reported

27-Oct-64 07:32 GMT Pacific Ocean 23°09'N 159°31'W Pacific
USS Falgout USN/ 1 Luminous object/ vapor trail/ Speed = 600 knots; flight at low altitude; picked up by radar/ Luminous vapor trail/ Missile like/ 205 deg Telex from USS Falgout to COMHAWSEAFRON MERINT

05-Nov-64 06:30 GMT Pacific Ocean 31°50'N 173°00'E Pacific USCG Cutter Matagorda Coast Guard/ 1 Bright object/ like satellite/ 60 deg arc in 8 sec; straight course/ Meteor like/ 5 min/ Elev. = 50 deg North Telex from Ship to CCGD 14th and USAF Chief of Staff & others MERINT
13-Nov-64 20:03 GMT Atlantic Ocean 65°32'N 28°08'W Atlantic
Atlantic Fleet; ship not reported USN/ 1 Bright celestial object/ 10 deg arc per min/ Satellite - Echo II /180 deg Telex from COMBARFORLANT to CNO, CINCLANTFLT, others CINCLANTFLT INST 03360.2C

17-Nov-64 20:05 GMT Atlantic Ocean 65°43'N 28°26'W Atlantic
Atlantic Fleet; ship not reported USN/ 1 Bright celestial object/ Satellite like/ 10 deg arc per min Satellite - Echo II /180 deg Telex from COMBARFORLANT to CNO, CINCLANTFLT, others CINCLANTFLT INST 03360.2C

18-Nov-64 03:30 GMT Pacific Ocean 30°00'N 140°00'W Pacific
USS Pontchartrain Coast Guard/ 1 Bright light/ Star like; satellite like/ Straight course/ High altitude/ Satellite - Echo II/ 165 deg/ Telex from USCGC Pontchartrain to COMDT COGARD, CINCNORAD, Others MERINT

19-Nov-64 11:00 GMT Pacific Ocean 34°55'N 164°05'E Pacific
USS Matagorda Coast Guard X/ Bright light /White/ flashing light/ Speed over 10K mph/level flight/ Unidentified/ 10 sec/ Elev. = 15 deg 20 deg/ Telex from USCGC Matagorda to CCGD 14th, Chief of Staff USAF, CINCNORAD, others MERINT

19-Nov-64 19:00 GMT Pacific Ocean 33°59'N 164°04'E Pacific
USS Matagorda Coast Guard/ 1 Bright object/ Satellite like / Straight course/ Elev. = 23 deg 140 deg /Telex from USCGC Matagorda to CCGD 14th, Chief of Staff USAF, CINCNORAD, others MERINT

20-Nov-64 01:12 GMT Middle of Puerto Rico but Radar caught by USS Gyatt 18°10'N 66°12'W Atlantic/ F-8C Pilot & USS Gyatt USN/ 1 Delta shape per pilot/ Size of jet fighter/ Dark (black or gray); light source emitting from tail of delta shape during periods of acceleration. No lights on board except what looked like after burner light. UFO made numerous course changes and then stayed in a parallel course. Pilot attempted to get close and UFO accelerated in 20 degree climb when within 5 miles. Performed no unusual maneuvers except extreme acceleration and deceleration at will, plus a very steep climb angle. Heat source; trail visible; no contrail/Aircraft/ 5 min/ 1 witness/ Radar Operator Initially/ NE; target made numerous course changes/ Telex from COMCARIBSEAFRON to CINCLANTFLT, others. Full report from CO, Utility Squadron Eight to

Commander Caribbean Sea Frontier regarding Pilot sighting. Full report from Commander Norfolk Test and Evaluation Detachment to FT regarding radar scope photos. CINCLANTFLT INST 03360.2c & OPNAV INST 3820.9

20-Nov-64 01:12 GMT Atlantic Ocean 20°41.5'N 68°34'W Atlantic
USS Gyatt USN/ 1 Radar detection only/ Radar / Aircraft /5 min/ 1 witness/ A/C Pilot Captured by A/C Captured by Radar

22-Nov-64 06:30 GMT Pacific Ocean 33°58.5'N 164°03.5'E Pacific
USS Matagorda Coast Guard/ 1 Bright object/ Satellite or star like/ Straight course/ Elev. = 34 deg South/ Telex from Matagorda to COFS USAF MERINT

24-Nov-64 04:20 GMT Pacific Ocean 34°00'N 140°00'W Pacific
USS Pontchartrain Coast Guard /1 Bright light /Star like/ Straight course/ Satellite - Echo I/SE Telex from USCGC Pontchartrain to COMDT COGARD, CINCNORAD, Others MERINT

28-Nov-64 14:36 GMT Pacific Ocean 30°00'N 140°00'W Pacific
USS Pontchartrain Coast Guard/ 1 Bright light/ Star like/ Straight course/ Satellite - Echo II Telex from USCGC Pontchartrain to COMDT COGARD, CINCNORAD, Others MERINT

28-Nov-64 15:50 GMT Pacific Ocean 33°55'N 164°10'E Pacific
USS Matagorda Coast Guard /1 Bright object/ Satellite like / Straight course/ Elev. = 10 deg 270 deg/ Telex from USCGC Matagorda to CCGD 14th, Chief of Staff USAF, CINCNORAD, others MERINT

29-Nov-64 06:25 GMT Pacific Ocean 34°20'N 163°48'E Pacific
USS Matagorda Coast Guard/ 1 Bright light/ Straight course/ Elev. = 25 deg 320 deg/ Telex

from USCGC Matagorda to Chief of Staff USAF, CCGD 14th, CTF 33, CINCNORAD, others MERINT

29-Nov-64 08:00 GMT Pacific Ocean 33°55'N 166°02'E Pacific
USS Matagorda Coast Guard /1 Bright light/ Straight course/ Satellite - Echo I / Elev. = 45 deg South/ Telex from USCGC Matagorda to Chief of Staff USAF, CCGD 14th, CTF 33, CINCNORAD, others MERINT

30-Nov-64 06:30 GMT Pacific Ocean 33°58'N 164°14'E Pacific
USS Matagorda Coast Guard/ 1 Bright object/ Satellite like/ Straight course/ Elev. = 30 deg 270 deg Telex from USCGC Matagorda to Chief of Staff USAF, CCGD 14th, CTF 33, CINCNORAD, others MERINT

30-Nov-64 08:45 GMT Pacific Ocean 33°59'N 164°82'E Pacific
USS Matagorda Coast Guard/ 1 Bright object/ Satellite like/ Straight course/ Elev. = 30 deg 170 deg/ Telex from USCGC Matagorda to Chief of Staff USAF, CCGD 14th, CTF 33, CINCNORAD, others MERINT

19-Dec-64 05:29 GMT Pacific Ocean 01°14'N 164°12'W Pacific
USS Charles Berry USN/ 1 Circular/ Pin Head/ White; alternating bright and dim; white when overhead/ Straight course/ Satellite like/ 6 min/ 4 witnesses/ Officer of the Deck, Junior Officer of Deck, Executive Officer, Quartermaster of Watch/ Elev. = 22 deg, then 90 deg, then 04 deg South/ Naval Message from USS Charles Berry to PACAF Base Command, CINCPACFLT, CNO, CTF 92, others CINCPACFLT INST 3820.3

10-Jan-65 07:15 GMT Atlantic Ocean 63°50'N 08°35'N Atlantic
Atlantic Fleet USN /4 Bright objects/ One object was blinking/ Straight course/ Elev. = 30

deg 170 deg/ Telex from COMBARFORLANT to CINCLANTFLT, CINCNORAD, others CINCLANTFLT INST 03360.2C

19-Feb-65 02:37 GMT Pacific Ocean 16°15'N 109°29'W Pacific
USS Serrano USN/ Spherical/ Pea/ Bright white/ Straight course/ No trails or exhaust/ Satellite/ 1 min/ 4 witnesses/ K. Weldon LTJG USNR Officer of the Deck; G.W. Zwirshitz ENS USNR Asst. Navigator; B.E. Albright QM3 USN Quartermaster of the Watch; J.A. Staacy SN USN Lookout/ Elev. = 65 deg 105 deg/ Telex from USAF So to ADC & others. Passing message from ship. OPNAV INST 3820.9 & CINCPACFLT INST 3820.3

20-Feb-65 03:45 GMT Atlantic Ocean 64°12'N 11°35'W Atlantic
Navy Ship not Named USN/ 1 Bright object/ Star like/ Straight course/ Satellite - Echo II/ 160 deg/ BB Index Card Only/ Not Reported

22-Mar-65 23:22 GMT Atlantic Ocean 64°36'N 29°13'W Atlantic
Atlantic Fleet USN/ 1 Bright light/ Bright as Vega; 1st Mag star/ Pale blue green/ Straight course Satellite - Echo II /5 min/ Elev. = 15 deg East /Telex from COMBARFORLANT to CNO, CINCLANTFLT, others CINCLANTFLT INST 03360.2C

11-May-65 05:35 GMT Pacific Ocean 30°00'N 140°00'W Pacific
USS Pontchatrain Coast Guard/ 1 Bright light/ Star like/ Straight course/ Satellite - Echo II/ Elev. = 50 deg 150 deg/ Telex from USCGC Pontchartrain to CMDT COGARD, CINC NORAD, others MERINT

02-Jul-65 00:15 GMT South of Capetown South Africa 34°24'S 23°41'E Atlantic
Antarctic research and resupply ship = MV R.S.A./ Civil/ 1 Round Small orange, Blue-white then orange before explosion/ Moving east;

then stationary for 1 sec and exploded/ 50 nm/ Meteor 5 sec/ 2 witnesses/ S.G. Stigant, Second Officer, K.T. McNish Master/ Elev. = 35 deg East/ Letter from Defense Attaché Office of the American Embassy in South Africa to FTD at Wright-Paterson AFB forwarding UFO sighting report. AFR 200-2

05-Aug-65 13:26 GMT Pacific Ocean 31°00'N 130°22.5'W Pacific
USS Isle Royale USN/ 1 Circular Dime/ Bluish white/ 1 Object appeared to turn straight up taking elliptical shape with decreasing minor axis until disappearance/ Contrails emerging from both sides of object: size twice the size of object/ Rocket/ 5 min/ 7 witnesses/Ruth (Navigator), White (OOD), Scott (JOOD), Jarvis (QM1), Robrecht (QM2), Eckhart (Messenger), Imes (Lookout) Elev. = 33 deg then 85 deg 343 deg Naval Message from USS Isle Royale to CINCPACFLT INST 3820.3

27-Aug-65 19:09 GMT Far East 31°09'N 136°54'E Pacific
USS George K. Mackenzie USN/ 1 Bright light /Pinpoint of light; 2nd Mag Star/ White/ Heading across celestial sphere/ Satellite/ 5 min/ 2 witnesses/ Lt JG Rossman and Lt JG Krakkfr/ Elev. = 60 deg to 40 deg 110 deg/ Naval Message from USS GK Mackenzie to CG Fifth AB Fuchu CINCPACFLT INST 3820.3

17-Sep-65 08:40 GMT Pacific Ocean 34°27'N 155°00'E Pacific
USCG Cutter Chautauqua Coast Guard /2 Bright light/ White/ Straight course / Satellites/
1-North East 2-East Naval Message from Air Div. Kunia Facility (Hawaii) to PACAF Hickam AFB Hawaii MERINT?

19-Sep-65 14:45 GMT Far East 34°20'N 163°45'E Pacific
Ocean Station Victor Coast Guard / 1 Bright light/ Alternating/ white, green, and red/

Stationary Sirius Star/ 19 min/ Elev. = 10 deg then 15 deg 120 deg/ Telex from OSV Victor to CINCPACFLT, COMWESTAREA, etc. then to COMHAWSEAFRON then to COFSUSAF MERINT

10-Oct-65 12:00 GMT Pacific Ocean; Off-Coast West of Portland Oregon 46°09'N 126°02'W Pacific/ Civilian Ship - Name not Reported/ Civil/ 2 Round bright objects/ Size of stars/ White/ Stationary until disappearance. Object rose rapidly, the larger of the two went North the other South/ 5 witnesses/ 1- North 2- South/ Ship reported sighting to West Port Coast Guard Station who passed report on to Naselle AF Station who called Wing Command Post McChord AFB Wash to ATIC/ Not Reported

17-Nov-65 15:03 GMT Pacific Ocean 28°57'N 178°49'W Pacific
USS Croatan USN /1 Bright light/ Bluish light/ whole horizon/ Celestial explosion/ Red contrails seen for 2 min/ Missile like/ 2 min/ Elev. = 42 deg/ Not Reported/ Telex from Air Div. Kunia Facility to Wright Patterson AFB with Merint report MERINT

29-Nov-65 14:40 GMT Pacific Ocean 30°00'N 140°00'W Pacific
USCGC Pontchatrain + Air Visual Coast Guard/ 1 Bright light/ Star like/ Surrounded by aurora like light/ Rising from Horizon/ Missile (Atlas D) / 240 deg Telex from CGC Pontchartrain to CINCNORAD, CSAF, COMWESTAREA, others, MERINT

29-Dec-65 16:31 GMT Atlantic Ocean 40°59'N 11°40'E Atlantic
USS Rich USN/ 2 Bright objects/ Similar to 1st Mag Stars/ White; 2nd object came 1 hr. later/ Moving rapidly in the sky, one after the other/ Satellites like/ 1- East 2- South Telex from USS Rich to NAVCOMNSTA ROTA Spain then to CONSIXTHFLT/ Not Reported

11-Jan-66 12:00 GMT Pacific Ocean 26°58'N 154°46'E Pacific
SS Morgantown Victory/ Civil/ 1 Cigar shape /200 ft at 400 ft height; a mile away Glowing object with a bright light at its head/ Straight course from horizon and then altered course, then hovered for 30 seconds (Object maneuvered approximately 180 deg about the vessel) Long fiery tail/ 1 mile off ship/ Satellite Decay (Cosmos 53) 5 min/ 3 witnesses/ Anderson (3rd Mate), Claunch (bow lookout), and Farba (helsman) From Horizon to 400 ft Elev./ At 1 Mile distance SE Letter from Commander USN Defense Forces Eastern Pacific to ATIC/ Not Reported

31-Jan-66 23:13 GMT Atlantic Ocean 50°08'N 31°01'W Atlantic
SS American Commander? / Civil/ 1 Bright light/ White/ Straight course/ Trail covered/ 15 deg of arc/ contrails/ On Horizon /258 deg/ Naval Message from NAF LAJES and COMEASTSEAFRON to CSAF and CINCNORAD MERINT

02-Mar-66 23:45 GMT Atlantic Ocean 18°28'N 66°07'W Atlantic
USCG Division 7 Coast Guard/ 1 UFO N/A Not Reported /Insufficient Data/ No Details/Telex from CCGD7 to CINCLANFLT/ Not Reported

27-Jun-66 04:00 Local Pacific Ocean 19°00'N 172°00'E Pacific
SS Mt. Vernon Victory /Civil X /1 Growing luminescent cloud /Size of golf ball at arm's length/ Flashing light at center of cloud/ Stationary cloud; white light moved southward/ Unidentified/ over 18 min/ 2 witnesses/ Radio Officer and Watch Officer Donald Rominger (Sorensen)/ Elev. = 10 deg to 40 deg/ Upward or toward the ship (West)/ Letter from witness directly to BB and NICAP/ Not Reported

20-Oct-66 02:00 GMT Pacific Ocean 32°00'N
117°20'W Pacific
USS Higbee USN /1 Round/ Size of 1st Mag Star
white / Straight course/ Satellite (Echo II)/
9 minutes/ 3 witnesses/ OOD, Watch
supervisor, other / Elev. = 30 deg then 15 deg
N/ Naval message from ship to Norton AFB
CINCPACFLT INST 3820.3

16-Feb-67 04:15 GMT Wake Island 31°28'N
130°32'W Pacific
USS Shearwater USN /1 object with 3 distinct
orange rays/ turning blue in color on final
burnout/ Straight course/ 2 witnesses/ No BB
Index Card/ Elev. = 40 deg to 20 deg from 220
deg to 175 deg/ Telex from ship to
COMWESTSEAFRON NAVY San Francisco then
to Western NORAD Region/ Not Reported

19-Apr-68 01:00 GMT Atlantic Ocean 24°14'N
83°32'W Atlantic
SS Louisiana Sulphur/ Civil/ 1 Flaming Object/
Yellow-orange/ Straight course; high speed
Object was breaking up/ Satellite Debris
(Cosmos 213) South Merint Telex from SS
Louisiana Sulphur to COMASWFORLANT, then
to CINC NORAD MERINT

23-Apr-68 06:21 GMT Pacific Ocean 23°24'N
166°41.5'W Pacific
USS Steinaker USN/ 1 Triangular Pea sized/
White; pulsating 90 times per minute/ Straight
course/ No trail/ 50K ft/ Probable Satellite/ 60
sec/ 3 witnesses/ OOD, BM3, and Watch
stander Elev. = 7 deg from 23 deg to 55 deg
/Telex from ship to CINCPACFLT then to ATIC
AFR 80-17 and CINCPACFLTIN ST P0054 4.10

Chapter 10

Military Bases of Interest

Atlantic Undersea Test and Evaluation Center (AUTEC)

Figure 35 By Wikited (talk) (Uploads) - Own work, Public Domain, https://en.wikipedia.org/w/index.php?curid=16739077

We are aware of The Atlantic Undersea Test and Evaluation Center (AUTEC), which is the American naval base on the Bahamas island of Andros. The U.S. Navy's Atlantic Undersea Test and Evaluation Center (AUTEC) is an instrumented laboratory that performs integrated three-dimensional hydrospace / aerospace trajectory measurements covering the entire spectrum of undersea simulated warfare: calibration, classifications, detection, and destruction. Its vital mission is to assist in establishing and maintaining naval ability of the United States through testing, evaluation, and underwater research.

Some researchers believed AUTEC might be an underwater "Area 51"- a place where secret research was being carried out on UFOs by the American government, and which, from time to time, was even visited by UFOs. It is located 177 miles southeast of West Palm Beach, Florida, at Andros Island and the Tongue of the Ocean, in Bahamas. The Andros Island AUTEC test facility-access must be obtained beforehand. It covers only one square mile on land, but actually comprises 1,670 square miles of the surrounding Caribbean. This ocean area is a steep-sided deep-water embayment 100 miles long and 20 miles wide, with depths more than a mile at the northern end. The Andros base has an ultra-top-secret caliber of security. In the waters off Andros Island, strange craft have been seen from time to time which not only resemble UFOs, but which display the same unbelievable swiftness of motion and execute the same incredibly sharp motion of turns.

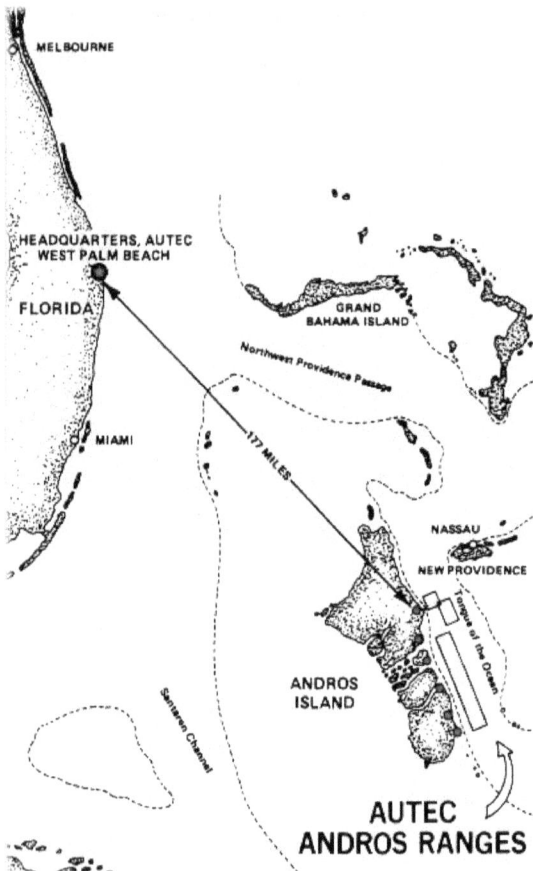

Figure 36 By U.S. Navy - U.S. Navy brochure showing location of AUTEC to job applicants., Public Domain, https://en.wikipedia.org/w/index.php?curid=21540803

Acoustic Research Detachment (ARD)

The Navy's Acoustic Research Detachment (ARD) at Lake Pend Oreille, Bayview, Idaho, is 375 miles from the Pacific Ocean and is thought to be the Navy's lower-key subsurface Area 51. Located in Lake Pend Oreille, it is a water based, smaller, and more outsider friendly Nellis Range Complex where new submarine and surface ship shapes and subsystems are tested in a sub-scale environment that closely mimics the ocean. At 1150 feet deep, this small base has supported every major submarine design development of the last seventy plus years and has been in existence since WWII. There are two "known" large- scale vessels which have been reported built on this base, but some say there are as many as ten in inventory. This research center is 375 miles inland, so one can only wonder what has become of seventy years of small- and large-scale submarines? The cost of these submarines to design is astronomical. Personal 2-person submarines come at a price starting at $80,000. Navy Submarines begin at $2.3 billion, with the price of the USS Navy Virginia-class attack submarine costing the U.S. navy $2.4 billion. I don't believe these proto-type submarines are just being sunk or dismantled, and because of the size, they most likely are not being trailered out. I "asked around" quizzing some of my sources, and the consensus through speculation is the existence of an underwater-underground 100-foot diameter tunnel beginning in Lake Pend Oreille leading out to the ocean 375 miles to the west. There is also a "local" legend of the existence of a "loch ness type" creature in the lake, but I will speculate that what is being seen and reported by witnesses is most likely a submarine periscope.

U. S. Navy's S.E. Alaska Acoustic Measurement Facility (SEAFAC)

Just to the north of Ketchikan, Alaska, near the Behm Canal, you will find the home of SEAFAC, the United States Navy's Southeast Alaska Acoustic Measurement Facility. Built in 1991, its existence was kept secret except from the locals, until recently. SEAFAC develops new technologies and equipment configurations which are tested aboard multi-billion-dollar American nuclear submarines and is additional where their acoustic signature is measured while at sea. The term "acoustic signature" refers to the noise something makes, in this case an American submarine, which is a critical concern while on maneuvers or in a battle situation. Advanced secret technology is used to develop quieter mechanical systems, pump-jets, and applying coatings on a submarine's hull, to isolate vibrations and sound waves. The shape and design of a submarine and how the water flows around it can be a major noise factor.

SEAFAC's focus is exclusively on increasing the submarine's chances of survivability. Behm Canal is a perfect area for an underwater acoustic test range because of its seclusion from the ocean, large amounts of vessel traffic, and ambient noise. Behm Canal also supplies a perfect basin setting because of its sooth tub-like contour which provides close to lab-like as possible acoustic conditions, which is and has been easily cordoned off when tests are underway.

Static and dynamic acoustic tests are conducted at the base using two acoustic arrays which are attached to the seafloor. They are placed at a depth of 1250 feet, which capture a submarine's audible signature as it transverses the canal back and forth while performing various tests. A pair of custom barges are also used as support for static tests

Figure 37 SEAFAC Photo: Wikipedia Public Domain

which can include an entire submarine or single component. Submarines are submerged while suspended on cables between two hydrophone arrays at various depths with the barges providing external power. This allows the submarine systems to be turned off while the testing occurs allowing to perform tests using various forms of sensor engagement technology.

Naval Undersea Warfare Center (NUWC)

The Naval Undersea Warfare Center (NUWC) is the United States Navy's "full-spectrum research, development, test and evaluation, engineering and fleet support center for submarines, autonomous underwater systems, and offensive and defensive weapons systems associated with undersea warfare." The Naval Sea Systems Command, NUWC is where one of the corporate laboratories, Division Newport is headquartered which is located in Newport, Rhode Island. There is also another subordinate facility, Division Keyport, which is located in Keyport, Washington. The Fox Island Facility and Gould Island Facility are also controlled by the NUWC which employs more than 4,400 civilian and military personnel, with a budget of over $1 billion.

The Naval Torpedo Station researched and tested underwater weaponry beginning before World War I and continued into World War II creating additional facilities on Rose Island, Fox Island and Gould Island. The station on Goat Island was reorganized in 1951, and for the next 15 years, was the Underwater Ordnance Station, and then the Underwater Weapons Research and Engineering Station. In 1970, the Underwater Sound Lab from New London, Conn., was combined with the Newport Facility to form the Naval Underwater Systems Center or NUSC, and then in 1992, the command was again reorganized as the current facility, Naval Undersea Warfare Center, Division Newport.

Yulin Naval Base, China

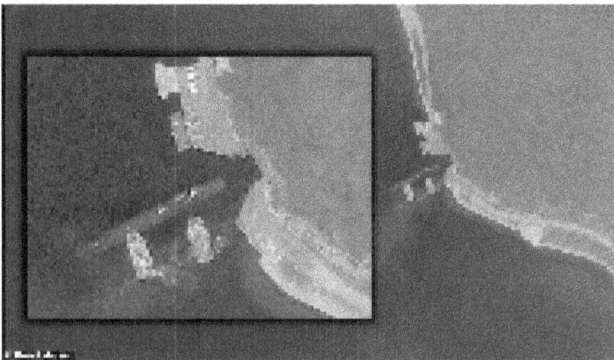

Figure 38 Ulin Naval Base Photo: Google Earth

The Yulin Naval Base is a naval base for nuclear submarines along the southern coast of Hainan Island, People's Republic of China. This underground-underwater base has been reported by several intelligence agencies. The images collected by the Federation of American Scientists (FAS) in February 2008 shows that China constructed a large-scale underground-underwater base for its naval forces. Wikipedia

This information is over a decade old and most certainly outdated, but significant none the less. We are also aware from CSIC/AMTI Digital Globe photos that China is building military facilities in the South China Sea. These facilities are being built on offshore waters claimed as territory by China for military purposes.

Russian Bases

Balaklava Submarine Base
The Russian Balaklava Submarine Base is located just off the Black Sea in the town of Balaklava, located on Balaklava Bay in the Ukraine. This base that dates back to the Cold War, was constructed in 1957 and comes with a nuclear shelter large enough to house 3000 people. The base which also included an armory, a channel, and a submarine repair center, was shut down in 1995 is currently open as a museum.

Yamantau Mountain
Near the city of Mezhgorye in the Ural Mountains, is said to be the location of a massive underground Russian base. The U.S. government became aware of a secret underground Russian base in the Yamantau Mountain in 1992, and the story was broke to the public in an article in The New York Times in 1996. Unofficially, it is believed that construction of the site began in the 1970s and the base is a massive 400 square miles.

Zhitkur
Zhitkur, referred to as Russia's Area 51, is rumored to be beneath Kapustin Yar, a Russian rocket research laboratory and launch site in Astrakhan Oblast. This base is said to be a place where "advanced technologies" are engineered and backengineered.

Arctic Trefoil Base

The "Arctic Shamrock," is a 14,000-square-mile, three-pointed base, and is now Russia's most northern permanent installation. The base will house 150 personnel and hosts multiple air defense units. Russia is taking a serious political stand by staking its claim for the resources beneath the Arctic. It has submitted a claim to the U.N. that 460,000 square miles of ocean floor should be considered its territory.
Reference: PATRICK REEVELL MOSCOW Apr 29, 2017, 10:12 ET

Inside the main building of the base is a five stories high atrium with fake plants and bright lights with the central space being supported by pillars. All the buildings are on stilts, which is a defense against the cold and the winter snow and are joined by walkways. A large antenna on the facility, appears to be a P-18 radar (NATO reporting name "Spoon Rest D," Russian designation 1РЛ131 "Терек"), an early warning system; reported range of 250 km.

English-language statement from the Russian MoD: "Military infrastructure of the Arctic base included numerous special constructions, control centers, storages for special and military hardware, and autonomous energy unit. As well as the living quarters, the complex includes an energy block consisting of a boiler, generating station and fuel cisterns, an automobile park consisting of a number of garage units for storing and servicing vehicles and special equipment, water pumping and treatment buildings, a recycling complex, storage for heating and lubrication materials, and a fueling point."

Northern Clover

The Russian Artic Base known as Northern Clover, opened in September 2015 and features a three-wing structure and a number

Figure 39 Russian Arctic Trefoil Base Photo:

of outbuildings joined by walkways. It offers various amenities, including volleyball, table tennis, and a sauna. According to Russian Defense Ministry statements, the base houses a tactical group of Russia's Northern Fleet, numbering over 250 people. It is designed to be self-sustaining in supplies and fuel for a year and a half. The base also includes what appears to be a garage for a coastal rocket division, and helipads for Mi-26 helicopters.
Conclusion:
"The Russian military has, understandably, maintained a degree of secrecy over its military complexes in the Arctic. However, Russia's construction workers have not kept to the same standards, revealing detailed plans of both areas. Thanks to the following social media users, the accidentally revealed details fill in some of the gaps left by the MoD's public relations exercises, illustrate a little more of the functioning of Russia's two newest Arctic bases."
Ben Nimmo is Senior Fellow for Information Defense at the Atlantic Council's Digital Forensic Research Lab (@DFRLab). Maks Czuperski is the Director of the DFRLab.

United States Antarctic Program

The United States has been studying the Antarctic and its interactions with the rest of the planet since 1956. The U.S. Antarctic Program employs staff investigators and supporting personnel with the goal of "supporting the Antarctic Treaty, fostering cooperative research with other nations, protecting the Antarctic environment, and developing measures to ensure only equitable and wise use of resources," The National Science Foundation funds and manages the program which is made up of 3,000 Americans yearly.

Figure 40 Mcmurdo from ob hill

There are three U.S. year-round research stations:

- **Ross Island (McMurdo Station)**
- **South Pole (Amundsen-Scott South Pole Station),** which is at geographic south and sits at the Earth's axis on a shifting continental ice sheet several miles thick. The South Pole supports projects ranging from cosmic observations to seismic and atmospheric studies. One astrophysical project at the South Pole is called the "IceCube", which is a one-cubic-kilometer international high-energy neutrino detector built in the clear ice, 1.25-2.5 kilometers below the South Pole Station.
- **Anvers Island (Palmer Station)** in the Antarctic Peninsula region

The three research goals for all stations:

- To understand the region and its ecosystems
- To understand its effects on (and responses to) global processes such as climate
- To use the region as a platform to study the upper atmosphere and space field science in Antarctica, is more expensive than in most places because of the remote location and extreme climate. Large, ski-equipped LC-130 airplanes, which are operated by the Air National Guard crews, provide air logistics, helicopters flown by contractors provide support for research teams and tracked or wheeled vehicles provide transport over land and snow. Small boats are also used in coastal areas.

A live webcam to the South Pole is available 24 hours at:
www.usap.gov/videoclipsandmaps/spwebcam

U.S. Submarine Bases:

military.wikia.org

The Naval Submarine Medical Research Laboratory (NSMRL) is located on the New London Submarine Base in Groton, Connecticut.

Naval Submarine Base Bangor, a former submarine base of the United States Navy, merged with Naval Station Bremerton into Naval Base Kitsap in 2004. In 1942, Base Bangor became a site for shipping ammunition to the "Pacific Theater of Operations" during World War II. The U.S. Naval ammunition magazine was established on this base in January 1945. From World War II, into the Korean War and through the Vietnam War, continuing until January 1973, the Bangor Annex remained as an active U.S. Navy Ammunition Depot responsible for shipping conventional weapons abroad. In 1973, the Bangor Base became home-port for the first squadron of Ohio-class Trident Fleet Ballistic Missile submarines, and on February 1, 1977, the Trident Submarine Base was officially activated. This base serves as the base for the U.S. Pacific Fleet.

Coco Solo was a United States Navy submarine base established in 1918 on the Atlantic Ocean side of the Panama Canal Zone, near Colón, Panama.

The Holy Loch is a sea loch in Argyll and Bute, Scotland. Robertson's Yard at Sandbank, a village on the loch, was a major wooden boat building company in the late 19th and early 20th centuries. During World War II, the loch was used as a submarine base, and from 1961–1992, it was used as a US Polaris nuclear submarine base. In 1992, the Holy Loch base was determined to be unnecessary following the demise of the Soviet Union and was closed.

Naval Submarine Base Kings Bay is a United States Navy Base located adjacent to the town of St. Mary's, Georgia, not far from Jacksonville, Florida. This Submarine Base is the U.S. Atlantic Fleet's home port for the U.S. Navy Fleet ballistic missile nuclear submarines which are armed with Trident missile nuclear weapons. The base covers about 16,000 acres of land, of which 4,000 acres are protected wetlands.

Naval Base Kitsap is a U.S. Navy base located on the Kitsap Peninsula in Washington state. It was founded in 2004 by merging the Naval Station Bremerton with Naval Submarine Base Bangor. Naval Base Kitsap serves as the home base for the Navy's fleet throughout West Puget Sound. It also provides base operating services, including support for surface ships, and Fleet Ballistic Missile and nuclear submarines. Naval Base Kitsap is the third-largest Navy base in the U.S. which features one of the U.S. Navy's "four nuclear shipyards, one of two nuclear-weapons facilities, the only West Coast dry dock capable of handling a Nimitz-class aircraft carrier and the Navy's largest fuel depot."

Naval Station Norfolk (IATA: NGU, ICAO: KNGU, FAA Location identifier: NGU) is located in Norfolk, Virginia, and supports naval forces in the United States Fleet Forces Command, operating in the Atlantic Ocean, Mediterranean Sea, and Indian Ocean. NS Norfolk, also known as the Norfolk Naval Base, occupies approximately four miles of waterfront space and seven miles of pier and wharf space of the Hampton Roads peninsula known as Sewell's Point. This Navy Submarine Base is the world's largest naval station, supporting 75 ships and 134 aircraft. Norfolk Base is also comprised of 14 piers and 11 aircraft hangars and houses the largest concentration of U.S. Navy forces, with its port services controlling more than 3,100 ships' movements of vessels annually. The Base's Air Operations conducts over 100,000 flight operations yearly, with an average of 275 flights per day. Over 150,000 passengers and 264,000 tons of mail and cargo depart on Air Mobility Command (AMC) aircraft and other AMC-chartered flights from the airfield's AMC Terminal annually.

Naval Station Pearl Harbor

Naval Station Pearl Harbor is a U. S. military facility adjacent to Honolulu, Hawaii and is the headquarters of the United States Pacific Fleet. The facility was merged with Hickam Air Force Base to form Joint Base Pearl Harbor-Hickam in 2010. This Honolulu Navy base is well known for the attack on Pearl Harbor by the Empire of Japan on Sunday, December 7, 1941, which brought the United States into World War II. Naval Station Pearl Harbor not only provides berthing and shore side support to surface ships and submarines but is also a maintenance and training base which can accommodate the largest ships in the fleet, to include dry dock services, and is now home to over 160 commands.

Naval Submarine Base New London

known as the "Home of the Submarine Force", founded in 1872, was the U. S. Navy's primary submarine base. The Navy Yard was first used for laying up inactive ships and later due to lack of funding, as a coaling station by the Atlantic Fleet small craft. It was located in the towns of Groton and Ledyard. By 1912, oil replaced coal in warships and again the Yard was scheduled for closure and the land relinquished by the Navy. The Navy Yard was spared permanent closure in 1912 by Congressman Edwin W. Higgins of Norwich, who was worried about the loss of Federal spending in the region. The Federal government spent over a million dollars on the

Yard in the next six years. On October 13, 1915, the monitor Ozark, a submarine tender, and four additional submarines arrived in Groton. The war in Europe and the Atlantic demanded the need for additional submarines and support craft so the facility was saved and listed as the Navy's first Submarine Base.

New Suffolk was the home to submarine USS Holland (SS-1), the first commissioned submarine in the U.S. Navy, along with six other Holland Torpedo Boat Company submarines which were based in New Suffolk between 1899 and 1905. Many believed this base to be the "First Submarine Base" in the United States.

Naval Base Point Loma is located in Point Loma, a neighborhood of San Diego, California. This base was established on October 1, 1998, when Navy facilities in the region were consolidated under Commander, Navy Region Southwest. The base consists of seven facilities which include a Submarine Base, Fleet Antisubmarine Warfare Training Center, Fleet Combat Training Center Pacific, Space and Naval Warfare System Command (SPAWAR), SPAWAR Systems Center, the Fleet Intelligence Command Pacific and Naval Consolidated Brig, Miramar which form a diverse and highly technical hub of naval facilities. The base population combined, is approximately 22,000 Navy and civilian personnel.

US Navy Submarine Base, Ordnance Island

Island, Ordnance Island, Bermuda, is located near the King's Square, in St. George. This base served as a Royal Army Ordnance Corps (RAOC) depot, supplying munitions to forts and batteries around the Colony. By WWII, the base was no longer useful in this role. From 1942 to 1945, the US Navy was allowed to use the Island as a submarine base.

Naval Base Ventura County (NBVC) is a premier naval installation composed of three operating facilities - Point Mugu, Port Hueneme and San Nicolas Island. Strategically located in a non-encroached coastal area of Southern California, NBVC is a key element in the DoD infrastructure because of its superior geographical location. NBVC supports approximately 80 tenant commands with a base population of more than 19,000 personnel. Tenant commands encompass an extremely diverse set of specialties that support both Fleet and Fighter, including three warfare centers: Naval Air Warfare Center Weapons Division, Naval Surface Warfare Center Port Hueneme Division and Naval Facilities Engineering and Expeditionary Warfare Center. NBVC is also home to deployable units, including the Pacific Seabees and the West Coast E-2C Hawkeyes. www.cnic.navy.mil

MANNED UNDERSEA STRUCTURES-THE ROCK-SITE CONCEPT

By, C.F. Austin Research Department
ABSTRACT: "Large undersea installations with a shirt-sleeve environment have existed under the continental shelves for many decades. The technology now exists, using off-the-shelf petroleum, mining, submarine, and nuclear equipment, to establish permanent manned installations within the sea floor that do not have any air umbilical or other connections with the land or water surface, yet maintain a normal one-atmosphere environment within. This presentation briefly reviews the past and present in-the-floor mineral industry. The methods presently practical for direct access to and from permanent in-the-floor installations are outlined, and the specific operations and types of tools indicated. Initial power requirements and cost estimates are included."
U.S. Naval Ordnance Test Station China Lake, California October 1966

MUFON archives

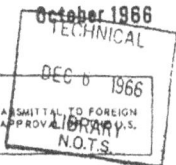

Figure 41 From the MUFON archives

"Feasibility of Manned In-Bottom Bases"

Research into and discussions of under-ocean and in-bottom military bases began decades ago. In 1968 the Stanford Research Institute discussed the construction of dozens of undersea bases. The study was titled "Feasibility of Manned In-Bottom Bases".

Wikileaks.org January 24, 2011

* "I've never heard of underwater military bases. Has anyone else done this? Brazil is evaluating a plan to build an underwater base to protect its offshore hydrocarbons deposits according to Jan. 13, 2011, reports. The

underwater platform, which would involve both government and naval, would be established at the edge of Brazil's maritime border. The base would allow Brazil to have deeper access to the offshore deposits and allow the military to better protect the country's oil assets."
* Not to my knowledge, and certainly not at the depths we're talking for Brazil's big offshore deposits.
* This is an oceanographic lab they plan to build, and the navy would be responsible to guard it. * Wow, interesting."

Nautilus Concept

Walter Koerschner, 1924-2017, was an artist and illustrator who was hired in the 1960's by the U.S. Navy. Koerschner was assigned a project for the United States Navy's Rock-Site Team, which was located at their China Lake, California Weapons Center. The project was to design and illustrate possible underwater submarine base facilities. Coincidentally, it was already rumored that the Navy was making plans to build large submarine bases beneath the ocean. The project known as the "Nautilus Concept" is capable of docking three submarines in one area, with an additional section that can dock multiple more. The square footage of the base would have to be significant taking into consideration the size of not only one submarine but multiple submarines. The main docking area would probably be over a thousand feet long, and more than a hundred feet in diameter, with additional space added for crew living quarters, the crew possibly numbering in the hundreds. Considering that this facility would be deep underwater in cold, dark, conditions including storms and water currents, completing this project would be very difficult and very expensive. In 1969, William B. McLean creator of the Sidewinder air-to-air missile and the former Director of the China Lake Naval Base, made a public comment to

John Newbauer, editor-in-chief of astronautics and aeronautics stating, "These manufactories and projects are already under construction."

In the fall of 2003, Koerschner made contact with Richard Sauder after hearing him being interviewed on Coast-to-Coast radio. Koerschner offered Sauder some very interesting, one-of-a-kind illustrations he had drawn for a government Navy project. Sauder gladly and gratefully accepted Koerschner's offer, accepting the drawings which were sent to him by mail. These illustrations are featured in Dr. Richard Sauder's book, "Hidden in Plain Sight".

Chapter 11

Suspected Underwater "Alien" Bases

The following witness reports are of events which lead us to believe the existence of "alien" life in the waters below. These reports are witness testimony of large craft seen below the water surface and of underwater "alien" encounters. They are testimony of unknown, unidentified "alien" craft leaving and entering a body of water as a resident would their home, and of witness reports of craft traveling underwater with ease at great rates of speed. These witness reports and numerous others not listed in these pages, are in themselves proof that "alien" life have taken up residence deep in our planet's oceans, seas, and lakes.

Richard Sauder, Ph.D. "Hidden in Plain Sight"

Conclusion:

- 1. There are many secret underground and probable undersea bases and tunnels
- 2. They can be impressively large and surprisingly deep
- 3. They contain very sophisticated technology
- 4. Many thousands of people are involved in constructing, operating, and maintaining these facilities
- 5. They are located all over the world
- 6. Aliens are probably involved in some, or maybe many, of the underground and undersea base
- 7. High-speed tube shuttle train systems probably do exist

- 8. We, the people of this planet, are being massively lied to by our so-called "leaders"

Alaska

Figure 42 Mount Hayes Alaska Photo: Scott Slone

Mount Hayes located in Alaska, is the highest mountain in the eastern Alaskan Range. The northeast face of the mountain rises to a steep 8,000 feet in only 2 miles which gives Mount Hayes the title of being the 51st "most topographically prominent peak" in the world. Bradford Washburn, Barbara Washburn, Benjamin Ferris, Sterling Hendricks, Henry Hall, and William Shand, were the first to climb the steep mountain face which was accomplished on August 1, 1941. Due to its extreme remoteness though, Mount Hayes is not frequented by mountain climbers due to the added access costs. As remote as Mount Hayes is reported to be and as difficult and costly to climb, where is all the aircraft traffic activity coming from? Numerous reports of unidentified craft "UFOs" have been reported in this mountainous area for decades. Some claim to have seen areas of the mountain open up for craft entry and then close behind with no trace of a door. Mount Hayes is currently considered a focal point for UFO activity and close scrutiny by researchers and UFO Investigators.

Chapter 5: The Alaska Triangle, lists several witness reports of unknown craft in that area. I have heard discussion of remote viewer attempts to access information concerning Mount Hayes, so I decided to access the newly released CIA files which are available on the "Black Vault" website, to see if anything of interest in the area was released. While in the "Black Vault", I discovered a government project referred to as "Project Sun Streak" from the 1980's which is described in the CIA files as follows:

- "Goal: Establish remote viewing as a reliable intelligence tool"
- "Psychoenergetics": Remote viewing (RV) ability to describe remote view areas or concealed data via unknown mental processes; ability to influence physical or biological systems via undefined physical mechanisms."
- "DIA's Motivation for pursuing this program: Establish whether a treaty exists, if so, what are its dimensions? Can the phenomenon be exploited?"

As I was searching for the released documents, I discovered something very interesting. I may be mistaken, but it seems someone was remote viewing Mount Hayes. The picture drawn and description by the remote viewer is pretty remarkable and seems very similar to Mount Hayes. Why would the CIA want to remote view a mountain in the middle of basically nowhere?

HAARP

Eielson AFB is 70 miles north of Mount Hayes with Fairbanks approximately 90 miles further to the northwest. The interesting detail about Mount Hayes is that it is almost directly in the center of the "Alaska Triangle", and in close proximity to three HAARP Stations; HAARP Gakona approximately 95 miles southeast, HAARP Fairbanks approximately 95 miles north, and HAARP Chatanika approximately 110 miles further north.

What is HAARP? Alaska's High-frequency Active Auroral Research Program, or HAARP, "is a scientific endeavor aimed at studying the properties and behavior of the ionosphere. "The ionosphere stretches roughly 50 to 400 miles above Earth's surface, right at the edge of space. Along with the neutral upper atmosphere, the ionosphere forms the boundary between Earth's lower atmosphere — where we live and breathe — and the vacuum of space." haarp.gi.alaska.edu

On August 11, 2015, the United States Air Force transferred its operation of the facility to the University of Alaska Fairbanks. HAARP is located on land which still belongs to the USAF, but a transfer of land ownership through legislative action is in the process. The UAF is currently working with the Army Corps of Engineers to complete the conveyance action.

High Frequency Active Auroral Research Program (HAARP)

Geophysical Institute
gi.alaska.edu

The National Defense Authorization Act will authorize approximately 1,158 acres and associated buildings to transfer to the University of Alaska.

Described by the University as being "the world's most capable high-power, high-frequency transmitter for study of the ionosphere", the function is the use of IRI in a controlled manner, allowing scientists to more accurately understand processes that occur continuously under the natural stimulation of the sun. This is done through the use of a high-power transmitter facility operating in a" High Frequency range". The IRI is used to "temporarily excite a limited area of the ionosphere for scientific study".

Could these experiments in the manipulation of the Earth's ionosphere be related to area UFO activity?

Anchorage and Prince William Sound

The Alaska Triangle has a very long history of UFO sightings. I was invited to appear on two seasons of the Travel Channels series "Alaska Triangle". I spent months researching the water areas, Ocean, Bays, and Islands and was surprised at how many UFO sightings have been reported over the years. It was also interesting to see how many ships have sunk in the area which seemed to be an unusually large number. While documenting these ships using Google Earth, I located some pretty interesting underwater anomalies in Prince William Sound. I charted and documented all of the listed boats and ships which have sunk in this area, but the Google Earth anomaly I found was not among them. I also found some interesting areas similar to what was found on Google Earth in Puerto Rico by Jorge Martin.

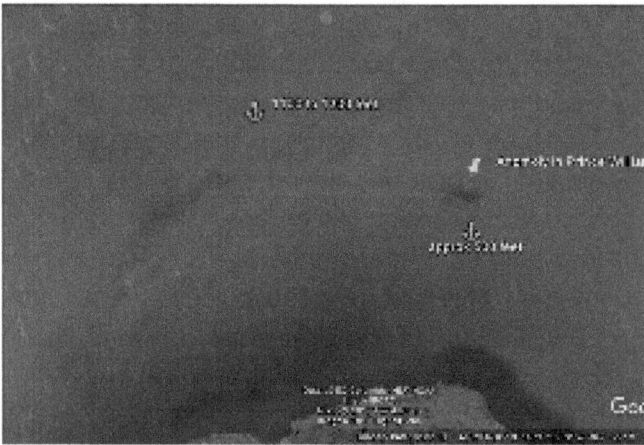

Figure 43 Prince William Sound
Photos: Google Earth

could be the result of colonization by different "alien-beings". An alien base doesn't necessarily need to be a "physical" base although Google Earth pictures show some strange underwater anomalies in the area. There is also the possibility of the area being accessed through a vortex or wormhole. I have also been informed of underwater bases with access openings appearing and disappearing. My source has informed me that these "alien" bases are in underwater mountainous areas accessible through a large "vortex" doorway that opens and closes in the mountain, leaving no trace.

Antarctic

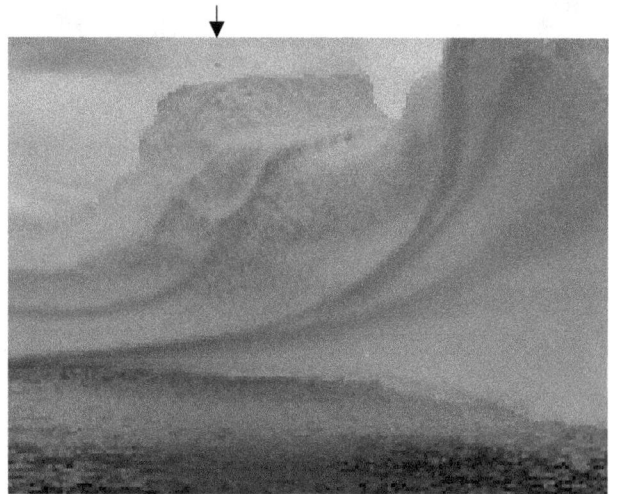

Figure 44 UFO in the Antartic Photo: WikiLeaks

Figure 45 Antarctic UFO Close up
 Photo: WikiLeaks

Even though there is a possibility of an "alien-being" base in the mountains to the north, there is still the possibility of a second "alien-being" base in and around the waters in the Anchorage and Prince William Sound area. More than one "alien" base in the same area

Figure 47 Photos courtesy of Wikileaks

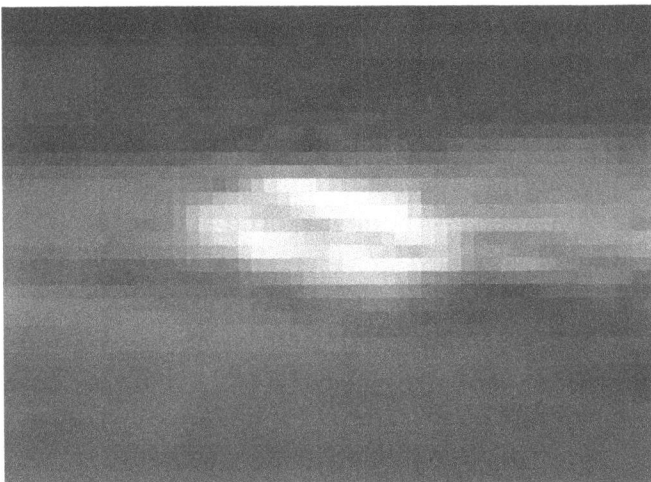

Figure 46 Close up

Julian Assange made comments a few years ago about his release of many controversial intelligence documents he was going to leak, stating that there would be UFO related materials included.

Supposedly, there was an all-out alert issued by Air Force Space Command after the emergence of a large fleet of airborne unidentified objects from the floor of the Southern Seas of the Antarctic on June 10, 2004. I could not find an exact WikiLeaks document to that effect, but I did come across 23 pictures released by Assange. In two of the released photos, I found something interesting in the sky which warrants further investigation.

Are these photos part of the UFO related materials spoke of by Assange?

Admiral Byrd

In 1929, Admiral Byrd launched the largest and most expensive expedition to Antarctica that had ever been attempted, (BAE I). The expedition had two goals; the first was to make the first flight to the South Pole, and the second was to take scientists in the fields of meteorology, geology, zoology, geography, and aerial photography, insuring the expedition's scientific purpose. In early 1929, Byrd set up a base camp which was named "Little America" located four miles from edge off the Ross Ice Shelf on the Bay of Whales. On November 29, 1929, the aircraft the "Floyd Bennett" took off towards the south pole, flown by pilot Bern Balchen, with navigator Admiral Byrd, photographer Captain Ashley McKinley, and radioman Harold June aboard. Barely clearing the Transantarctic Mountains, the plane and crew made its way safely to the south pole. The Admiral took several sextant observations to confirm the fact that they had flown a circle around the imaginary South Pole point, and then Byrd dropped an American flag weighed down with

Figure 48 Antarctic Expedition Courtesy of Joe Fex, original photographer unknown

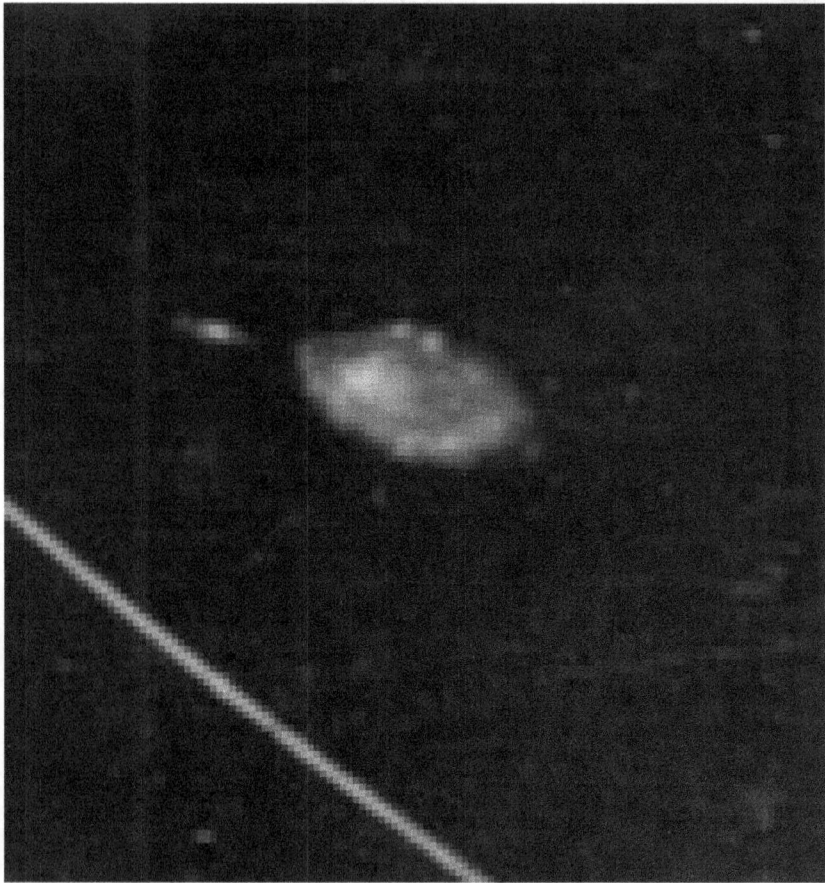

Figure 49 Possible UFO over Antarctic Base Close up from previous photo
Photo courtesy Joe Fex

a stone from his good friend Floyd Bennett's grave. I have a friend who gave me a copy of an original picture he obtained in an auction. The original picture which I saw, was small and viewed through a stereoscope. The picture he purchased along with others are believed to have been taken in the Antarctic by polar explorer, Admiral Byrd or one of his associates. Close observation of the picture shows several very interesting objects in the sky near the "Little America" radio-antennas. admiralbyrd.com

November 07, 2016
John Kerry, after departure from a trip to Antarctica, discussed a new agreement with the United Nations supposedly about global warming. However, there exists a thirty-five-year ban on private ships from traveling to Antarctica without advanced authorized permission.

Is this ban designed to cover-up secrets from the past and what is currently being conducted in lakes under the Antarctic? Are there hidden human military and or alien force bases?

Figure 50 Base in Antartica
Photo: admiralbyrd.com

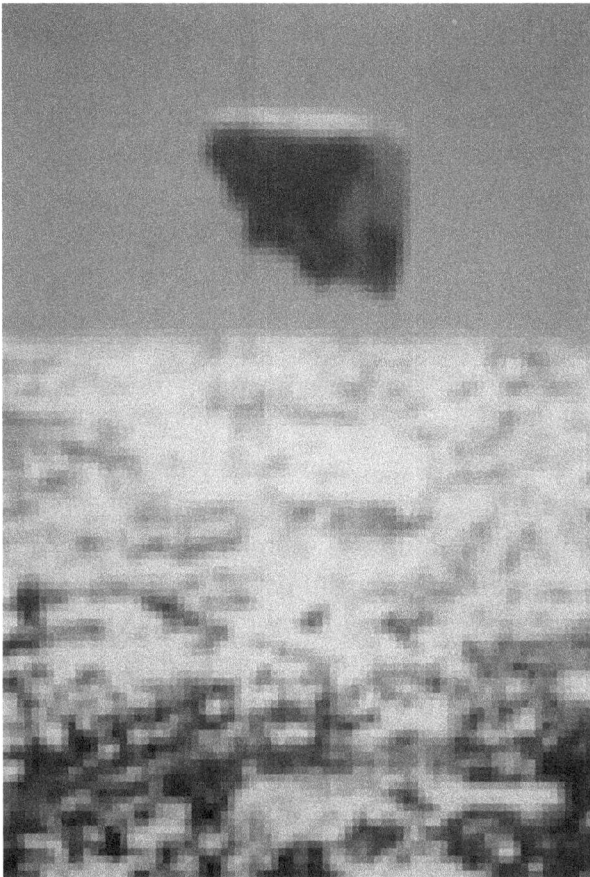

Figure 51 UFO photo taken from USS Trepang SSN674 Submarine
Photo courtesy express

Guantanamo Bay

MUFON Case #74794 1968-69
Witness Interview Statement: Direct transcript testimony:

"I have kept this secret for decades and do not wish to go to my grave with the things that we all witnessed in Cuba. I was a Marine and stationed at Guantanamo Bay. I mostly worked fence duty as part of the security forces there. I remember the first time I saw what could only be described as a UFO, I was amazed and didn't expect to see anything like it again. But I did. We all did. There were things going on every night it seemed. Some of these lights would fly overhead less than 300 feet! Most of what we saw were lights, but they were surrounded by a cloud-like substance or a haze. You know like what you see on very hot days. I did have a memory recently that once or twice I could make out a boomerang shape. Most of the sightings had to have been large crafts because of the cloudy/haze around the lights that I cannot verify were always connected with just one craft every time. Most were no less than 50 feet to no more than 100 feet across. It's difficult to identify the exact size even while seeing it. There seemed to be a small red light trailing them most times too! That's how we knew that was the end of the craft sightings. These objects were seen while working perimeter duty or the fence line of the base.

One night, I was on guard duty at the main gate. This is where most traffic in and out would be. It was about 7 p.m. and dark at this time, I stepped out of the guard shack and looked outside of the gate to the Cuban Guard house and saw immediately a white cloud-like hazy formation with blueish white lights or a baby blue color if you will, pulsating in the middle of this cloud substance. I asked the other Marine guard with me if he saw this too. It was now rising up over the Cuban Guard shack and moving towards our position. Just then, we heard an urgent call from the Marine SGT. in the watch/observation tower yelling for us to get the hell out of there! It was too late as this cloud was directly over us at this time and we didn't move. It was completely silent. We heard nothing! I believe we stood there for several minutes before hearing again the SGT. from the tower ordering us to pull back and leave the shack. Basically, abandon a strategic post and head back to our barracks. This was also a very strange order. The barracks were about 200 feet away on a hill and I remember quite clearly seeing a crew of Base Intelligence personnel now filming this object. I saw them film for about 3 hours. 3 solid hours! Then the cloud with the lights started moving down the fence line in a western direction. After what appeared to be about 1/4 mile, the object stopped and shot straight up like a bullet until we couldn't see it anymore. This was very fast!

We also witnessed on the south side of the island at another guard post, {post 13}, many, many times, UFOs coming into and out of the ocean. These were mostly blue or whitish lights and large. Again, it would hard for me to put an exact size on them. But they would move around while out of the water. When they would re-enter the ocean, they appeared to slow down and dim that I knew was the result of them descending. We would watch these for an hour or two. Sometimes a saucer shape would be seen but these were quick. I know for a fact that we were well aware of them. We witnessed Jets chasing them over the ocean.

I was never frightened by what I witnessed and quite frankly enjoyed seeing them. I was fascinated more than anything, but some were frightened by what they saw.
We were never offered a brief or explanation, and I talked with many Marines about the events throughout my tour of 13 months. This was something known to be happening on this base at least during 1968 and 1969. Before my transfer was completed, I was brought to HQ on Base and a Major sat down and told me to not talk about what I saw, and I did sign a secrecy document that lasted 10 years."

"I decided to come forward now because I am 70 years old and do not wish to keep this secret with me. The 10 years are long gone."
Location: Naval Base, Guantanamo Bay Cuba
Evidence/Investigation: Yes. This case does require an "investigation" and verification of military service and duty station assignment. A Secrecy Agreement was signed. His service record and time in Gitmo were verified, authentic.
Witness Credibility: High

Santa Catalina Island, California

The area of water between the coastlines of Southern California and Santa Catalina Island, 20 miles offshore to the southwest, following south down the coastline to the far edges of Mexico, has been a "hot spot" for UFO sightings since the 1940's. The state of California has more UFO sighting reports to MUFON than other state in the U.S. The majority of these are along it's 1,100 (10 degrees of latitude) of Pacific Ocean coastline. The beach at Point Dume at night has been a popular "UFO sky watch" sight for many years with UFO enthusiasts searching the sky above the ocean for the multi-colored UFOs that some claim," sink under the water at times." MUFON field investigators receive many witness reports of UFOs traversing the Palos Verdes Peninsula in Southern California and UFOs coming out of and hovering over the water in the San Pedro Channel. Reports of UFO sightings have come from Catalina Island since the 1940s.

The witness reports come from fishermen, tourists, scuba divers, pilots, police officers and even military personnel. Sighting reports are of strange unknown craft hovering over the water or passing over-head at a high rate of speed. Witnesses have reported unidentified underwater lights with some witnesses even reporting beans of light coming from the water. One of these "beams" took down a small aircraft killing the passenger and severely injuring the pilot. The most famous California UFO coastline sightings was the "Battle Los Angeles" of February 25, 1942 (2:36am) which is mentioned in Chapter: 3.

Five years later in the early fall of 1947, steamers going into and out of San Francisco Bay encountered an "undersea mountain" that appeared and disappeared in various locations in the bay. The mysterious mass was reported by several ships many referring to it as a "reef" or "submarine mountain" that appeared overnight. The same mass was reported by another ship, which described what they saw as "a large mass under water, off the Golden Gate." Soon after these many accounts, the mass mysteriously disappeared.

Following the 1947 incidents sightings came regularly, in 1951, 1954, 1955, 1957, 1962, 1964, 1970, 1980, 1990, 1991, 1993, 2004 the list goes on, most from the Santa Catalina channel. The sightings involve many highly credible witnesses including lifeguards, security guards, law enforcement officials, military officers, and countless citizens.

Topanga Canyon, just 17 miles east of Malibu, has been a UFO hotspot for more than 50 years. On the night of June 14, 1992, it exceeded sky watcher's expectations when hundreds of unidentified craft were seen in the canyon area. UFO reports were filed from seventeen adult witnesses who reported seeing UFOs on that evening. One couple, who lives high on a ridge overlooking the ocean, reported seeing approximately 200 craft rising up, one by one, from behind the ridge east of them, then moving to various locations in the canyon. Stated by one witness; "You know when you watch something for a while, you can figure out where they're coming from. After watching them, you got the feeling that they were going all over this area from that certain spot right there." The "certain spot" described by the witness, is the same location where numerous underwater UFOs have been seen in the past. The 200 plus craft were coming from below, either underground or underwater. From their canyon vantage point, many witnesses were not sure.

Malibu, California Coastline

There is an underwater anomaly located off of the Malibu, California coastline at Point Dume. This plateau structure was discovered by Maxwell, Dale Romero, and Jimmy Church host of Fade to Black on Dark Matter Radio on May 12, 2014. The structure is located at 34*00'02N 119*01'41W 1.35 miles x 2.45 miles wide 6.66 miles from the shore with the entrance between the support pillars being 2745 feet wide and 630 feet tall. What appears to be the ceiling is 500 feet thick. Many believe this anomaly to be a secret underwater "alien" base, and others believe this to be a "Dume thrust" which is a natural phenomenon.

Figure 52 Malibu anomaly Photo: Google Earth

Public released statement by David Schwartz of the U.S. geological Survey:
This is interpreted as a thrust fault," Schwartz said, "meaning one side of the crust moves up over the other -- and what we're looking at is interpreted as being the surface expression of this Dume thrust, which is part of a large fault system in Southern California."

In 2017 MUFON and a TV series titled "The Lowe Files", starring Rob Lowe, sent an ROV (remotely operated vehicle) which is an unoccupied, highly maneuverable, underwater camera, deep down in the dark Pacific Ocean to investigate the anomaly. Due to rough sea conditions, and poor visibility, the findings were inconclusive.

Numerous UFO sightings have been reported from this area of beach in southern California beginning back in the 1940s. Malibu Beach with its magnificent surf and large waves, has been a favorite spot for surfers for decades. Numerous UFO sightings have been reported by surfers many with multiple witnesses. Fishermen offshore from Malibu beach have also sighted unexplained craft or anomalous lights coming out of or going into nearby waters.

Preston Dennett, UFO researcher and author of more than 20 books on paranormal and supernatural phenomena, states that he believed in the alien base theory for years. "I was convinced there was something there before this Malibu anomaly was publicized. I've collected probably 50 or 100 reports of people who have seen UFOs going in and out of the water there. My problem is that the Google images are coming out different, depending on what viewpoint you're looking at this thing from. Some show a tunnel. Some don't." Dennett remains optimistic about the anomaly being something extraterrestrial. "I'm very intrigued by the possibility. I'm convinced there's something down there."

I agree with Preston, and if given the opportunity to catch a ride on a mini sub to take a look, Preston is welcome to ride along. In Chapter: 3, I listed many of the sightings I have come across over the years in the Catalina Island area. I lived in southern California back when I was a teenager and still have family members in Orange County. I've been to Catalina Island several times and always enjoy the ferry ride out, I don't get seasick. Schools of as many as fifty or more Dolphins follow the ferry and often times a whale tail or two will be visible. The trip to

Avalon from New Port Beach is only a bit over an hour-long ride and money well spent. For more details on Santa Catalina Island UFO sightings a very good book was written by Preston Dennett called:
Undersea UFO Base: An In-Depth Investigation of USOs in the Santa Catalina Channel

Lake Baikal, Russia

Figure 53 Lake Baikal Photo: Inaki LLM

Lake Baikal has a depth of 4,911 plus feet. The lake is credited with holding over twenty percent of the planet's fresh water and harboring more indigenous species of flora and fauna than any other lake on the face of the Earth; including numerous fish species found nowhere else. People often report UFO encounters in Russia's Lake Baikal, the deepest freshwater body in the world. Fishermen tell of powerful lights coming from the deep and objects flying up from the water.

In 1982, a group of Russian military divers training underwater at Lake Baikal, spotted a group of humanoid creatures dressed in silvery suits. The encounter happened at a deep diver depth of 165 feet. The divers attempted to catch the strange creatures unsuccessfully resulting in three of the seven men dying, while four others being severely injured. The encounter occurred as follows; Major-General V. Demyanko, commander of the Military Diver Service of the Engineer Forces of the Ministry of Defense, USSR spoke to local officers of an extraordinary event that had occurred during training exercises in the Trans-Baikal and West Siberian military regions. During their military training dives, the frogmen (scuba divers) had encountered mysterious underwater swimmers, very human-like, but huge in size described as being almost three meters tall (10 feet). These swimmers seemed to be wearing tight-fitting silvery suits in icy-cold temperature waters. Seen at a depth of fifty meters, these "swimmers" wore no scuba diving equipment of any kind, only sphere-like helmets concealing their heads. The local military commander made the decision to attempt capture one of the creatures so a special group of seven divers, under the command of an officer, had been dispatched for the task at hand. As the frogmen attempted to cover one of the creatures with a net, the entire dive team was thrown out of the deep waters to the surface by a very powerful force.

Surfacing from depth at a high rate of speed and not allowing for a steading pace of ascent and safety stops to avoid decompression sickness can be deadly. The condition of decompression sickness or aeroembolism occurs when a formation of nitrogen bubbles accumulates in the blood and other body tissues. Divers release these nitrogen bubbles from depth through a series of decompression stops before reaching the water's surface. Treatment for decompression sickness is immediate confinement under decompression conditions in a pressure chamber. Several pressure chambers were available in the military region, but only one was in working condition and could accommodate no more

than two divers at a time. Four of the divers were forced into the chamber resulting in the death of three of them which included the leader of the group. The remaining three divers were just left to die.

A major-general was rushed to the Issik Kul to warn the local military, in particular Officer Shteynberg, against similar "devil-may-care" actions. The Issik Kul Lake is shallower than Baikal Lake, but the death of the Frogmen ascending from depth in such a manor was a risk not worth repeating. Reports had come in from Issik Kul Lake of similar creatures. The major-general obviously knew more than he was letting on. The staff headquarters of the Turkmenistan military region had received an order from the Commander-in-Chief of the Land Forces detailing, analyzing, and reprimanding those involved in the Baikal Lake events. This was supplemented by an information bulletin from the headquarters of the Engineer Forces of the Ministry of Defense, USSR. listing numerous deep-water lakes where there had been registered sightings of "anomalous phenomena: appearances of underwater creatures analogous to the Baikal type, descent and surfacing of giant discs and spheres, powerful luminescence emanating from the deep, etc." These documents had high level classifications and were "for the eyes" of a very limited circle of military officers with the purpose being "to prevent unnecessary encounters."

Late in April 2009, astronauts aboard the ISS (International Space Station) observed a strange circular area of thinned ice in the southern end of Lake Baikal in southern Siberia. There was another, very similar circle near the center of the lake above a submarine ridge that bisects the lake. Both circles were visible from April 20, 2009, into June. The origin of their presence was both mysterious and unexplained. How the circular areas were formed on the ice in Lake Baikal was never explained.

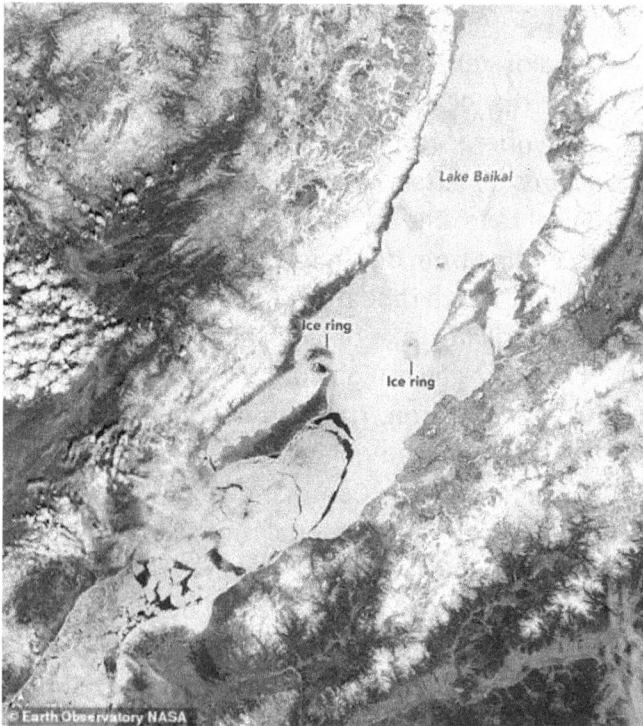

Figure 54 Lake Baikal   Photo: ISS Public Domain

SETKA

One of the most active areas for UFO sightings in Russia is located in the Voronezh region in the town of Borisoglebsk. Borisoglebsk, with a population of approximately 65,000 people, is located on the left bank of the Vorona River. This town is also home to secret Soviet UFO research under the name of SETKA-AN. SETKA-AN in part, was formed by the USSR Academy of Sciences due to increase anomalous atmospheric phenomenon reports between 1977 and 1978. A second program used for the same purpose as the first, SETKA-MO was formed by the Soviet Ministry of Defense. By the 1980s, unidentified objects and craft were reported seen in the area by the military. These objects ranged in sizes as small as an orange to as much as 600 feet across, traveling up to 400 mph, at levels ranging from ground-level to more than twelve miles high.

Giant Swimmers

Unidentified very large humanoid creatures have been reported seen swimming deep in Soviet waters for decades. An article was published in 1984 by Yekaterina Vorontsova of a mission in the Baltic Sea in which unusual looking creatures were seen by the crew of a Soviet Submarine. The creatures were described as "human-like" swimmers dressed in silver suits, approximately ten feet tall, swimming at ease in a dangerous depth of 1,300 feet.

Another report came from B.Borovikov, hunter of Black Sea sharks, who described seeing while scuba diving in the Anapa area at a Tech depth of over 250 feet deep, "giant beings ascending from below". He described the beings as "milky-white, with humanoid faces, and tails that resembled that of a fish." The lead creature, saw Borovikov, stopped with its eyes bulging, focused on him in an erring stare, and was then followed by two additional creatures from the deep of the same species who also stopped and starred. The lead creature waved its hand towards Borovikov, then the three creatures moved forward together towards him then stopped a short distance away. After just a brief moment, the three creatures turned around and together, descended back into the deep dark waters of the Black Sea.

An additional incident in the 1990s was reported by D. Povaliyayev who was hang gliding over Kavgolovo located not too far from Leningrad. As the hang glider was descending, he noticed something odd swimming in one of the many lakes below him. He described "three giant fish-like creatures", which as he descended lower, took on the appearance of "swimmers in silvery costumes".

Kolyma

In the Arctic in the northeast corner of Siberia, is a region referred to as the "Kolyma River" which is bordered by the East Siberian Sea, the Artic Ocean in the north, and the Sea of Okhotsk to the south. In a book by Mikhail Gershtein, "Secrets of UFO and Aliens", Gershtein describes UFO sighting reports from KGB head Pyotr Pavlov, who presided over the section of Yakutsk from the late 1950s to the

Figure 55
Photo credit: Almiro Barauna

early 1960s. So many reports of "UFOs" from the Kolyma River close to the Arctic came across Pavlov's desk one summer that the Ministry of Defense was called to the area. In the sky above the town of Chersky, located at the mouth of the Kolyma River, five to seven disk-shaped objects were seen hovering at high altitude appearing in a straight line and departing minutes later in the same manner.

Another call was made to the KGB reporting a UFO landing on the shores of the Lena River in the Arctic Circle. Several witnesses described the object as "being an unknown object with glowing portholes silently landing on the snow." The area of snow briefly turned reddish color, the object omitted several red flashes, ascended silently, and disappeared. Another incident involved the crew of an aircraft that witnessed a disc-shaped craft appear over the Nizhniye Kresti settlement on

Kolyma. The disc-shaped craft was described as being a bit smaller than the size of the moon. The craft hovered over the southern area of the sky just above the horizon for three days. Appearing at approximately 15:30 each day, the craft would appear for approximately two hours, then move towards the west, ascend sharply, and disappear. First thought would be that the object was a planet or star, but an aircraft pilot flying at an altitude of approximately 23,000 feet, was able to see the craft as it moved slowly from the east to the west. The pilot described the object as "a disc, pearly-white, with no visible antennae or fixtures, and had pulsating edges." The disc-shaped craft appeared in the area for three consecutive days, left, and never returned.

Puerto Rico

La Poza Del Sur

The area of La Poza del Sur is located between the south of Vieques and the southeastern coastline of Naguabo. Very "strange" stars have been reported by local fishermen in this area, which they state, "jump and move around in the sky in different directions". After a brief time, these "stars" suddenly disappear and at the same moment, a popping sound is reportedly heard. Strange objects have also been seen zigzagging across the sky disappearing into the distance. There of La Poza del Sur is described by local fishermen as being an "undersea abyss" being one of the best fishing sites in the area.

Unexplainably to the locals, this area has an unusually long presence of a large U.S. Navy research ship which stays anchored at sea just south of Cerro Ventana-Pirata and the Playa Grande sector. The navy research ship is equipped with a huge crane and seems to be looking for something in an area of the Caribbean which falls to a depth of thousands of feet deep. Several personnel are aboard the ship, most wearing what appear to be silver plated overalls similar to some type of hazmat suit. In one instance, several fishermen surrounded the research ship hoping for some answers, but the ship's crew scurried around the ship's deck removing their silver suits and turning the anchored ship to block their view. Dolphins have also been seen by the local fishermen being places in the water from the research ship by means of a hammock and retrieved a few minutes later in the same manor. Undersea scanning has also been seen done from this ship and one other which occasionally joins in research with the first. The scanning was done "inch by Inch" across the entire southern coastline from east to west. Locals believe the U.S. Government's motivation was the knowledge of an alien presence in the region.

Hoya Del Sur

Carlos Ventura a leader of the Vieques' fishermen, told of an unusual event which occurred on at least two occasions while he was fishing the Hoya del Sur. Ventura threw down the fishing net tied to some rope using his boats sonar to register the depth of the bottom which registered at 2,230 feet. He threw the net expecting to hit bottom but "nothing, they continued falling beyond the 2,300 feet mark". Ventura remarked that "it was as if there was no bottom at all down there." He let out more rope until it reached a depth of 3,300 feet but there was still no bottom. Ventura speculated his rope entering an undetected underwater cave, but the sonar indicated that there was a surface down there, with a solid bottom. Ventura exclaimed, "I used the sonar equipment again and an image appeared showing there was ample bottom at a depth of 2,300 feet, but underneath that bottom or surface or whatever it was, there was an open space, and then you could see the image of another bottom, of another surface

farther down! To me there was no logical explanation".
Explosions and strange noises" which seem to come up from the sea bottom, have also been heard by Ventura and other local fishermen. Similar sounds were also reported in the mid-1980s by local fishermen in the southwestern section of Puerto Rico.

Cartagena Lagoon

Another area in Puerto Rico filled with mysterious sightings of unidentified craft and strange looking creatures is Cartagena Lagoon located just north of Cabo Rojo. Cartagena Lagoon has a long history of strange creatures, unidentified sounds, and people disappearing never to be seen or heard from again. But something more identifiable began on June 1, 1987, just below the Lagoon with an unidentified blast and earthquake to follow. Several local residents in the area witnessed a tsunami near the Cabo Roja Lighthouse. The blast and tremor were later found to have occurred 81,000 feet below Cartagena Lagoon. Several fishermen who were in the Sea south of La Parguera reported finding "strange particles" floating on top of the water and later upon coming out of the water, the fishermen began to develop a rash, itchy skin, and a burning sensation. Additionally, several of the fishermen were sick for several additional days experiencing muscle pains, fever, headaches, nausea, and hair loss. It was reported that a young girl also died shortly after experiencing the same symptoms as the fishermen.

Local Journalist Jorge Martin Interviews Witnesses

- In 1975 Professor Armstrong, Mayaguez University, was out looking through her telescope one evening, when she noticed an object almost outside the atmosphere, that she described as "cigar-shaped" with several smaller objects coming out of it. The objects were descending into the ocean below in the vicinity near Mararita Cay in El Beril. The professor also claimed that she had seen craft flying into the Lagoon on many occasions from 1956 to well into the 1960s. The craft usually came from the direction of Guanica, which she described as being "metallic, silvery crafts with a dome at their top which was also metallic, with many windows around the dome. She claimed she could on occasion see people, being, through the windows. Her reports of these incidents fell on death ears when reported to the local newspaper.

- Another witness to Cartagena Lagoon events was an elderly woman Mrs. Garrastazu whose family owned a piece of property near the lagoon. She stated that back in the summer of 1945, some U.S. military personnel accompanied by some scientists, contacted her father who was a police lieutenant, and asked him to take them by boat into Cartagena Lagoon. When they reached the center of the lake, the mem lowered a very long cable into the water but could not find the bottom. She also recalled an incident in 1921, when her uncle while our riding a horse one night, saw a huge red ball of fire come out of the lagoon. As he tried to shoot at the object, he and his horse became temporarily paralyzed.

Guayanilla 1995

In the summer of 1995, two pilots, one of whom was a member of the Puerto Rico Police Department, flew over an incredibly large object just south of Guayanilla in southern Puerto Rico. Officer Martin, the pilot, reported

to Journalist Jorge Martin, that they had flown out from Cabo Rojo on a heading towards Mercedita Airport in the city of Ponce. The day was a sunny afternoon, 3:00 p.m., no clouds described as a "beautiful sunny day". Officer Martin stated "At the time we were flying south of Guayanilla over the sea, and all of a sudden we saw that thing submerged just below the water's surface. It was an object similar to an enormous flying saucer." The object the two men saw was not very deep underwater, in fact, it was so shallow the men could make out distinct details. "It was round and sort of metallic with an opaque gray color all over it. It had four floors or sections, one on top of the other, and each one smaller than the previous one, like a cake. There were many square windows in each section, all around it, and on the last one, at the top, it had a big metallic dome of approximately 400 feet in diameter." Curiosity got the best of them so the two decided to circle back at a lower altitude for a better look. As they flew over the object, they realized the enormity of whatever they were seeing. The underwater object measured at least a mile in diameter, taking them nearly seven minutes to fly completely around it. At that lower altitude, they could see no activity inside or out, so they decided to continue their trip to Mercedita Airport. Upon arrival, the two men filed a report with the local authorities who informed them that they had received several similar reports from the same region.

Could this have been an underwater alien base in the process of relocation?

Puffin Island, Wales

Puffin Island, Wales is an area where many mysterious unknown lights have been seen leaving and entering the sea. Ufologists Phil Hoyle who has investigated this strange phenomenon, believes that the area may be an underwater alien base. Hoyle states that many of the sightings in early 1974 were of not just lights, but solid objects that were seen leaving the sea near the island. He also stated that he has heard and read reports from numerous witnesses about sightings at Puffin Island. Hoyle states that all of the witnesses tell the same story and describe the same type of phenomenon. Witnesses in the area claim that they have been abducted by alien beings and that their abductors were humanoid looking. The abductees state that they were told by the "aliens" that they came from a base under the sea near Puffin Island. There is an ancient legend of an underwater kingdom near Puffin Island called Cantre'r Gwaelod. Some witnesses believe that these humanoids are their descendants. Cantre'r Gwaelod (definition 'The Lowland Hundred') was supposedly between Ramsey Island and Bardsey Island known today as Cardigan Bay, which is west of Wales, UK. Legend states that Cantre'r Gwaelod extended approximately 105 feet west of the current shoreline into the bay. A legendary king named Gwyddno Garanhir, was said to have ruled over Cantre'r Gwaelod during the sixth century. The legend tells of the land being submerged under water when Mererid, a priestess of a fairy well, allowed the water to overflow.

Solomon Islands

A current suspected UFO base is referred to as "Mount Dragon UFO Base" which according to the locals, is the residence of alien "life-forms". The Solomon Islands entrance, leads to a subterranean area from a waterfall-lake at approximately 2,500 feet high on the western side of a three-mile-long valley, which is approximately five miles from the coast. Unidentified craft (UFOs) are seen regularly, sometimes nightly by local residents. As many as several hundred UFOs have been seen by the local Guadalcanal people who refer to these flying 'balls-of-light' UFOs as "Dragon Snakes". The residents of Solomon Island have seen UFOs for over one hundred years. The

locals tell tales of people being killed, injured, or have gone missing due to the UFO activity and the "alien" life-form residents who inhabit the mountain area. The Solomon Islands Prime Ministers "Sir Alan Kemakeza's" (current) and "Ezekiel Alebua" (past) are said to be aware of the Waterfall Lake UFO Base entrance and reportedly have seen numerous unidentified craft.

A more accurate location of the "alien" base known as Waterfall Lake or "Mount Dragon UFO Base", is in south Small Malaita. The locals claim you only have to head north up a passage with Small Malaita on your right, which is approximately two miles from Affiou and approximately half a mile up the jungle mountain, to find the alien base entrance. UFOs have been reported flying low over the jungle, have surfaced from out of the sea near fishermen, and have landed near villages. The alien beings are described as "white or black, giant, human-like people" who are wearing strange grey uniforms. The alien craft are described as "flat string-ray type or "triangular" shaped UFOs with big round lights underneath them that hum as they go along."

Spaceships Landed off Anglesey Coast

Claims UFO Expert - Wales Crash Info
by Geraint Jones North Wales Chronicle in January 2011

ALIEN spaceships under the sea might sound like the stuff of fiction, but a UFO researcher claims that an incident off the shores of Anglesey could be the best evidence yet of an infamous North Wales close encounter. However, UFO expert Russ Kellett said that a coastguard document backed his theory that one of three-alien craft found off the coast of Anglesey crashed into Lake Bala and caused the Berwyn Mountain incident January 23, 1974. He said: "This information that I have unleashed is a sort of Holy Grail. The Ministry of Defense released documents a year ago and there was nothing about military activity. "The researcher, who has investigated the UFOs for 23 years, said that the Marine Coastguard document revealed that a military exercise called Operation Photoflash was underway from Liverpool Bay to the North Wales coast on the night of the Berwyn Incident. Russ believes that the operation used depth charges to locate one alien craft at Puffin Island, another off the Anglesey coast and a further ship near Bangor. The spaceships then shot out of the sea and into the air. Russ said that a group of men discovered one of the silver flying saucers while the aliens emerged from one of the crashed ships on a roadside near Llandrillo. He said: "They were slim humanoid creatures, about five ft six inches tall, with grey skin and wore jumpsuits with belt kits on them. "Russ added that the men watched the solders load the aliens into a vehicle and drive away. The researched believed that the MoD had not released the information to prevent public panic. He said: "You've got something that travels about under the sea without detection and can fly. What a piece of engineering that is." "People are going to get really worried, because this is what I call super technology." Russ also claims to have several fragments of one of the craft. Documents released by the MoD last year support the theory that a meteor shower caused the bright lights and earthquake experienced by residents near the Berwyn Incident's epicenter at Bala Lake.   END

Vladimir Azhazha Russian Ufologist

Statement: "I think about underwater bases and say: why not? Nothing should be discarded. Skepticism is the easiest way: believe nothing, do nothing. People rarely visit great depths. So, it's very important to analyze what they encounter there." Vladimir Azhazha is one of the most prominent personalities in

the history of Russian UFO and USO research. He attended M.V. Frunze Higher Naval School, served in the Soviet Navy, and later participated in oceanographic research. Azhazha also served as the head of the underwater exploration expedition aboard the Soviet "Severyanka" submarine designing many of its features before becoming a USO investigator.

Abductees Taken to Underwater Bases

Betty Hill

Kathleen Marden, niece to Betty Hill, states in the book "Captured" The Betty and Barney Hill Experience (written by Kathleen Marden and Stanton Friedman) of a dream Betty Hill spoke of to her niece Kathleen during a regression session. The dream was of a nightmare Betty had about water, a lake, and a shoreline. Betty Hill who along with husband Barney, were abducted from New Hampshire on the evening of September 19 to 20, 1961. Betty dreams of this in association with "alien" presence but cannot remember details only the statement "it's a big orange ball and it's glowing and glowing and it's rolling just like a ball. It must be, I don't know, water? Do they go underwater? It goes down and then there is a dip, and then, zoom."

Linda Porter

Linda Porter claimed to be abducted by aliens at age 15 and taken to an underwater alien base off the coast of California near Santa Barbara in Covina in 1963. "I was supposedly taken to an underground base off the coast of California. For some reason I was led to believe it was in the Santa Barbara area. If you were standing on sand at the bottom of the ocean where this place was at, all you could see is what looks like a silver submarine conning tower rising up out of the sand. And it would probably be the height of a two-three story building. I was told this tower thing was camouflaged by an electronic net of some kind that renders it invisible. And they also have something around it that seems to repel people and fish for some reason. Inside the building the floors and the walls and the ceiling were all a silver-gray color. There was a lot of light. But the doors, and there seemed to be doors all over the place, were brightly colored. They were either bright red or bright blue or bright yellow. And all over each door, was some kind of writing that looked like hieroglyphic or Arabic writing or something like that." There was a praying mantis being and light-filled room. I remember the being slowly coming around the corner and facing me. He stood very still and simply waited, as if he knew how frightening he appeared to me. Eventually he began to talk to me, although as yet, I do not have any memory of what was said. He seemed to possess a great deal of dignity and gave the impression of being quite old. After he finished saying whatever he had to say, I turned around and he walked me to this room (filled with dense light). The memory ends with me about to enter the room and the thought, 'Translated into the Light.' Whatever that means, it has to do with what happened."

Avalon Harbor, Catalina Island, California

Two eleven-year-old boys experienced an episode of missing time in 1967, while on their parents' boat in Avalon Harbor, Catalina Island. Paul Nelson (alias) years later, underwent a hypnotic session and recalls an abduction that he and his friends experienced as children. Recalled under hypnosis, the boys were taken to an underground base where they were examined by praying-mantis-type ETs. Nelson (alias) recalls; "I was taken into a round-walled room. It seemed to me more underground than it did onboard a ship. The walls had kind of a rock-like facet to them...rock-like walls

rather than craft-type walls. It gave the impression that I was in a cavern, rather than a ship. it was more of an underground feeling." After the abduction and examination, the boys were returned to their boat with no conscious memory of the abduction.

Filiberto Cardenas

On January 3, 1979, Filiberto Cardenas claimed to be abducted from a highway outside of Hialeah, Florida while with his family. His car had broken down on a rural road where he and his friend Fernando Marti were attempting to work on the engine. He was taken to and shown an underwater base that aliens maintain near Miami. Cardenas awoke seated onboard a UFO where he saw three strange figures. He was then taken to a smaller ship where he saw the beach approaching, and then the UFO plunged into the sea. Cardenas could see a tunnel which the ship entered emerging in a place that was completely dry. He was welcomed by a human looking figure who said he was from Earth and had long worked with the alien beings. Cardenas had a conversation with the being, a quick trip to what is commonly accepted as being an undersea alien city and was then returned to a pasture near where he had been abducted.

(Hopkins, 1996)

As documented in the book "Witnessed" by Bud Hopkins, forty-one-year-old Linda Cortile (alias) later identified as Linda Napolitano is abducted out of her New York City apartment building through a window on November 30, 1989, a little after 3:00 a.m. The mother of two floats out from an apartment building window high above Manhattan's FDR Drive. Accompanied by three alien figures all suspended in a blue light beam, she is lifted into a large reddish-orange glowing UFO. The UFO then moves across the sky towards the Brooklyn Bridge.

Two government agents (Richard and Dan), who were near the Brooklyn Bridge early that morning, came forward and agreed precisely with Linda's account of the abduction. The two men were bodyguards of a senior United Nations diplomat who was visiting Manhattan and would eventually be identified as Javier Perez de Cuellar. According to his two bodyguards, the diplomat was "visibly shaken" while viewing the event. The three men not only witnessed the abduction, but the plight of a woman being floated through the air accompanied by three entities. They watched as the abductee Linda, was being floated and accompanied on a short trip to a massive hovering flying craft. One of the men described the passing of the craft as follows; "When it flew over us, I could feel the hair on my head, arms, and legs stand straight up, and it wasn't from fear. The static cling was incredible, the electricity was tremendous. I wasn't amused at all". The craft then plunged into the river not far from Pier 17, behind the Brooklyn Bridge where the UFO would come to rest.

As stated by agent Richard:

"There was an oval-shaped object hovering over the top of the apartment building two or three blocks up from where we sat. We didn't know where it came from. It happened too fast. Its lights turned from a bright reddish orange to a whitish blue coming out of the bottom. Green lights rotated round the edge of the saucer. A little girl or woman wearing a white gown sailed out of the window in a fetal position - and then stood in mid-air in this beam of light. I could see three of the ugliest creatures I ever saw. I don't know what they were. They weren't human." Their heads were out of proportion, very large heads with no hair. Those buggers were escorting her into the craft. My partner screamed, 'We have got to get them.' We tried to get out of the car but couldn't. After the woman was escorted in, the oval turned reddish orange again and whisked off."

Richard stated to Hopkins in a letter:

"Mr. Hopkins, the oval never came up from under the river. It's possible that it could have, after we drove away 45 minutes later. We would have stayed longer but we couldn't ignore our radio call any longer. The guilt is brutal, more so than the fear we felt when we witnessed this terrible encounter. The guilt has lingered into today and we find it difficult living with ourselves. My partner and I have been debating for fourteen or fifteen months if we should seek her out. We know the building and we know which window she came out of. Perhaps she was just a figment of our imagination. If she isn't, is she alive and well? We have to know."

Under hypnosis:

Linda could see through a large window on the craft that she was underwater. She described seeing a murky water bottom which see assumed was the East River. The lights of the UFO brightened the darken night water enough for her to see garbage and a soft-drink bottle floating by. Linda claimed that she was abducted by the so-called "greys," who floated her from a closed bedroom window into a hovering UFO. Linda's experience was at first, an experience of "lost memory", although she could recall bits and pieces of the abduction. She remembered the actual kidnapping and the room where she was examined, but how she was transported was totally lost to her.

Linda's own words:

"I'm standing up on nothing. And they take me out all the way up, way above the building. Ooh, I hope I don't fall. The UFO opens up almost like a clam and then I'm inside. I see benches similar to regular benches. And they're bringing me down a hallway. Doors open like sliding doors. Inside are all these lights and buttons and a big, long table. I don't want to get up on that table. They get me on the table anyway. They start saying things to me and I'm yelling. I think they were trying to tell me to be quiet because he put his hand over my mouth."

Bud Hopkins not only interviewed several other witnesses to the event but additionally found a "metallic object" implanted in Linda's body. All the details of this incredible event can be read about in the book "Witnessed" by Bud Hopkins.

Kim Carlsberg

As recounted in her autobiography "Beyond My Wildest Dreams", Kim Carlsberg, recalls her abductions in the early 1990s, exact dates not recalled, where she experienced a series of UFO abductions from her Malibu home. On each abduction experience she is examined by grey or praying-mantis-type ETs. On August 30, 1992, Carlsberg also remembers being taken to a large underground complex where she saw many other abductees and ETs of various types. "I woke up in a lobby where many humans were milling around. They reminded me of patients waiting to endure their turns in a dentist's chair." She also recalls sitting in a large "auditorium" with many other abductees where she was told by the ETs that she was being "prepared for something."

Chapter 12

"Blue Planet Project"

The following information was referenced from the document titled "Blue Planet Project" which is believed to be the personal notes and scientific dairy of a scientist who was contracted by the government over several years to visit all crash sites, interrogate captured "Alien Life Forms" and analyze all data gathered from that endeavor. This individual was discovered to have kept and maintained such personal notes that he is currently in hiding out of this country. It is believed that his involvement in these investigations span over a thirty-three-year period. He was soon discovered and immediately went into hiding in 1990. At this time, to the best of the data available to "Blue Planet Project", there are at least 160 species or races of Aliens from different galaxies, stars, and planets that governments and the private sector have encountered.

THE INVISIBLE GOVERNMENT

President Truman signed the National Security ACT on July 26,1947 and immediately named Secretary of the Navy Forrestal as the first Secretary of Defense who along others were sworn in September 17,1947. The ACT established "under the National Security Council" a Central Intelligence Agency headed at that time by "Director", Rear Admiral Roscoe Hillenkoeter, provided a comprehensive program for future Security of the United States. The ACT created the NSC to advise the President with respect to the integration of domestic, foreign, and military policies relating to the National Security, with

the special duty to "Access and Appraise the objectives, commitments, and Risks. The funds for the CIA were hidden in the annual appropriations for other agencies. Today, collectively, U.S. Intelligence Operations which is almost totally clothed in secrecy, cost more than $1 Billion Dollars annually.

OPERATION MAJESTIC 12-NSC, MJ-12 SPECIAL STUDIES PROJECT

MJ-12 is said to be a TOP SECRET Research and Development, Intelligence Operation established by President Truman on September 24,1947. This "Committee" was set up inside the NSC and in 1954, President Eisenhower signed the Secret Executive Order, "Order Number 54-12". The NSC called this group the "54-12" committee "which gave the President responsibility of approving all 'Black' covert projects." This committee has undergone several changes over the years, and currently the "40 COMMITTEE" which is represented as XXXX "the double - double cross." It is described as the "Directorate" of the NSC. The "40 COMMITTEE" has access to advanced technology and teams to cover-up, "the cover-ups." In the past, this committee was headed by Dr. Henry Kissinger (code name: 'The Overseer'). He gave William Colby permission to commission Howard Hughes, Summa Corporation, to build a submarine and a special craft or salvage vessel. The salvage vessel called "Glomar Explorer", was equipped with refrigeration capacity for up to, unexplainably, one hundred bodies. In early May 1988, President Ronald Reagan said he often wonders what would happen if the Earth was Invaded by a "Power from Outer Space". It has also been reported that all Presidents are briefed on UFO developments and "Alien Visitors, by Planetary Intelligence group #40 (PI-40).

OPERATION MAJORITY

Operation Majority is the name of the operation responsible for every aspect of the project and all consequence of Alien presence on Earth. Majesty was listed as the code word for the President of the United States for communications concerning this information.

PROJECT GRUDGE

PROJECT GRUDGE contains 16 volumes of documented information collected from the beginning of the United States investigation of the Unidentified Flying Objects (UFOs) and Identified Alien Crafts. The project was funded by the CIA, and money from the illicit drug trade. Participation in the illegal drug trade was justified in that it would identify and eliminate the weak elements of our society. The purpose of PROJECT GRUDGE was to collect all scientific, technological, medical and intelligence information from UFO and IAC sightings as well as contacts with Alien Life Forms. This orderly file of collected information has been used to advance the United States Air Force Space Program, which President Trump revealed as "Space Force" which he designated as a new branch of the military on December 20, 2019.

JASON SOCIETY (SCHOLARS)

President Eisenhower commissioned a secret society known as the Jason Society under the leadership of the Director of Central Intelligence, Allen Welsh Dulles, Dr. Zbigniew Brzezinski, President of the Trilateral Commission from 1973 until 1976 and Dr. Henry Kissinger, leader of the scientific effort, to sift through all the facts, evidence,

technology, lies and deceptions and find the truth of the Alien question. The society was made up of thirty-two of the most prominent men in the U.S.

MJ-12

MJ-12 is the name of the secret control group inside the Jason Society. The top twelve members of the thirty-two members of the Jason Society were designated the MJ-12. MJ-12 has total control of everything and are designated by the code J-1, J-2, J-3, etc. all the way through the members of the Jason Society. The director of Central Intelligence was appointed J-1 and is the Director of the MJ-12 group. MJ-12 is responsible only to the President of the United States.

Original MJ-12 Members
1. Secretary James Forrestal
2. Admiral Roscoe H. Hillenkoetter
3. General Nathan P. Twining
4. General Hoyt S. Vandenburg
5. General Robert M. Montaque
6. Dr. Vanevar Bush
7. Dr. Detlev Bronk
8. Dr. Jerome Hunsaker
9. Dr. Donald Menzel
10. Dr. Lloyd V. Berkner
11. Mr. Sidney W. Souers
12. Mr. Gordon Gray

The MJ-12 group has been an existing group since it was created with new members replacing others that died. It was originally organized by General George C. Marshall in July of 1947 to study the Roswell-Magdalena UFO crash recovery and debris. The "ROBERTSON PANEL", was activated by Admiral Hillenkoetter,(Director of the CIA) and designed to monitor civilian UFO study groups that were appearing all over the country. He joined NICAP (National Investigations Committee on Aerial Phenomena) in 1956 to act as the MJ-12 "Mole", along with his team of other covert experts. The goal was to steer NICAP in any direction they wanted, with misinformation. This put the "Flying Saucer Program" under complete control of MJ-12 with the physical evidence safely hidden away by General Marshal.

MAJI (MAJORITY AGENCY FOR JOINT INTELLIGENCE)

All information, disinformation, and intelligence are gathered and evaluated by this Agency. This agency is responsible for all disinformation and operates in conjunction with the CIA, NSA, DIA, and The Office of Naval Intelligence. This very powerful organization as well as all Alien projects, are under its control. MAJI is responsible only to MJ-12. MAJIC is the security classification and clearance of all "Alien" connected material, projects, and information. MAJIC means MAJI controlled.

ALIEN COVER UP IN THE U.S.

In the year 1947, which was two years after the U.S. set off the first nuclear explosion, came the "Cap-Mantell" episode, which was the first incident of a military confrontation with extra-terrestrials that resulted in the death of a military pilot. In 1952, the nation's capital was overflown by a series of disks. It was these events which led to the involvement of the United States Security Forces (CIA, DIA, NSA, & FBI) to try to keep the situation under control until they could come up with a "game plan". The members of the group MJ-12 were designated as the personnel with the task of dealing with the "alien" situation.

ROSWELL

Within six months of the July 2, 1947 crash in Roswell, NM., and the finding and retrieval of a second crashed UFO at San Augustine Flats near Magdelena, New Mexico on July 3, 1947, a reorganization of agencies and shuffling of people took place. The main goal behind the original "Security Lid", and the very reason for

its organization, was the analysis and attempted back engineering of the technologies of the "alien" flying discs. This project was turned over to The Research and Development Board (R&DB), Air Force Research and Development (AFRD), The Office of Naval Research (ONR,) The CIA Office of Scientific Intelligence (CIA-OSI), and The NSA Office of Scientific Intelligence (NSA-OSI). No single one of these organizations were to know the entire event story, which was kept compartmented
on a "need to know" basis. The NSA was created to protect the secret of the recovered flying discs, and eventually got complete control over all communication intelligence. This complete control allows the NSA to monitor any individual through mail, telephone, text, email, and computers, by monitoring private and personal communications as they desire.

SPECIAL PROJECTS

In December of 1947, **PROJECT SIGN** was created as a means of acquiring as much information as possible about UFOs, their performance characteristics and their purposes. In order to preserve security, liaison between **PROJECT SIGN** and **MJ-12** was limited to two individuals within the intelligence division of the Air Material Command. Their role was to pass along pertinent information through channels. **PROJECT SIGN** evolved **PROJECT GRUDGE** in December 1948, and **PROJECT GRUDGE** had an overt civilian counterpart named **PROJECT BLUE BOOK**, headed by J. Allen Hynek with only the "Safe" reports being passed to **PROJECT BLUE BOOK**. **PROJECT GRUDGE** contains sixteen volumes of documented information collected from the very beginning of the United States Investigation on Unidentified Flying Objects. **PROJECT AQUARIUS** was originally **PROJECT SIGN**, established in 1953 by President Eisenhower, under control of the CIA and MAJI but in 1960 the project's name was changed to **PROJECT AQUARIUS** and funded by CIA confidential funds. The purpose of **PROJECT AQUARIUS** was to collect all scientific, technological, medical, intelligence information, and contact with "alien life forms" information, from UFO and IAC sightings. These files of collected information have been used to advance the United States Air Force Space Program, which is not NASA. **PROJECT AQUARIUS's** mission was to compile the history of Alien presence and their interaction with homo sapiens on Earth for the last 25,000 years and culminating with the Basque people (PAIS BASCO) who live in the mountainous country on the border of France and Spain and the Assyrians (or Syrians, originally from the Star Sirius). **PROJECT SNOWBIRD** was established in 1954 with its mission to develop, using conventional technology, a "Flying Saucer" type craft for the public. This project was successful when a craft was built and flown in front of the PRESS which was used to explain UFO sightings and divert the public's attention. The **"MAJIC PROJECTS"** or **"SIGMA"**, was the project first established for communications with the Aliens and is still responsible for communications. **"PLATO"** is the project responsible for "diplomatic relations" with the Aliens which secured a formal treaty (illegal under the U.S. Constitution) with the Aliens.

TERMS OF THE "ALIEN-HUMAN TREATY" ESTABLISHED:

The Aliens would give us, "our Government" technology and would not interfere in our history. In return, we, "our Government" agreed to keep their presence on Earth a secret, not to interfere in any way with their actions, and to allow them to abduct humans and animals. The Aliens agreed to furnish MJ-12 with a list of abductees on a periodic basis for Governmental control of their experiments with the abductees.

1. **AQUARIUS** is the project which compiled the history of the Alien presence and interaction on Earth and the "HOMO SAPIENS".

2. **GARNET** is the project responsible for control of all information and documents regarding the Alien subjects and accountability of their information and documents.

3. **PLUTO** is a project responsible for evaluating all UFO and IAC information pertaining to Space technology.

4. **POUNCE** project was formed to recover all downed and/or crashed craft and Aliens. This project provided cover stories and operations to mask the true endeavor, whenever necessary. Covers which have been used were crashed experimental Aircraft, Construction, Mining, etc. This project has been successful and is ongoing today.

5. **NRO** is the National Recon Organization based at Fort Carson, Colorado. It's responsible for security on all Alien or Alien Spacecraft connected to the projects.

6. **DELTA** is the designation for the specific arm of NRO which is specially trained and tasked with security of all MAJIC projects. It's a security team and task force from NRO specially trained to provide Alien tasked projects and LUNA security (also has the CODE NAME: "MEN IN BLACK"). This project is still ongoing.

7. **BLUE TEAM** is the first project responsible for reaction and recovery of downed crashed Alien craft and

Aliens. This was a U.S. Air Force Material Command project.

8. **SIGN** is the second project responsible for collection of Intelligence and determining whether Alien presence constituted a threat to the U.S. National Security. SIGN absorbed the BLUE TEAM project which was a U.S. Air Force and CIA project.

9. **REDLIGHT** was the project to test fly recovered Alien craft. This project was postponed after every attempt resulted in the destruction of the craft and death of the pilots. This project was carried out at AREA 51, Groom Lake, (Dreamland) in Nevada. Project Redlight was resumed in 1972 and has been partially successful. UFO sightings of craft accompanied by Black Helicopters are project Redlight assets. This project in now ongoing at Area 51 in Nevada but probably moved seven miles away to area S-14.

10. **SNOWBIRD** was established as a cover for project Redlight A "Flying Saucer" type craft which was built using conventional technology. It was unveiled to the PRESS and flown in public on several occasions. The purpose was to explain accidental sightings or disclosures of Redlight as having been the Snowbird crafts. This was a very successful disinformation operation and was only activated when needed. This deception has not been used for many years and is currently in mothballs, until it is needed again.

11. **BLUE BOOK** was a U.S. Air Force, UFO, and Alien Intelligence collection and disinformation project. This project was terminated, and its collected information and duties were absorbed by project Aquarius. A classified report named "Grudge-Blue Book, Report Number 13" is the only significant information derived from the project and is unavailable to the public.

MAJIC's CONTINGENCY PLANS

In 1949, MJ-12 developed a plan of contingency called MJ-1949-04P/78 that was to make allowance for public disclosure of some data if necessary. PLAN A was a delayed release of information by use of MAJESTIC-12 in the form of disinformation to delay and confuse the release of information should anyone get close to the truth. PLAN B should the information become public, or should the aliens take over, called for a public announcement that a terrorist group had entered the United States with an atomic weapon. It would be announced that the terrorists planned to detonate the weapon in a major city. Martial Law would be declared and all persons with implants would be activated by the Aliens. Those individuals would be rounded up by MAJIC along with all dissidents and would be placed into concentration camps. The PRESS, the Radio and TV would be nationalized and controlled. Anyone attempting to resist would be arrested or killed.

MAJIC's SECRET WEAPONS AGAINST THE ALIENS

GABRIEL was a project to develop a High Frequency pulsed sound generating weapon. It was said that this weapon would be effective against the Alien crafts and their Beam weapons. **JOSHUA** was a project to develop a Low Frequency pulsed sound generating weapon and was tested at White Sands Proving Grounds. This weapon was developed between 1975 and 1978 and assembled in Anaheim, California. **JOSHUA** was a long-horn-shaped device connected to a computer and amplifiers and described as being able to totally level any man- made structure from a distance of two miles. **EXCALIBUR** was developed to destroy the alien underground bases. It is a nuclear missile capable of penetrating 1,000 meters of packed soil, such as that found in New Mexico, causing no operational damage. I'm sure other major weapons to use against Aliens have been developed but no data for these was available at the time the" Blue Planet Project" document was released to the public.

DEFINITIONS FROM GRUDGE-BLUE BOOK, REPORT NUMBER 13

1. **EBE** is the name or designation given to the live Alien captured at the 1947 Roswell, New Mexico crash. He died in captivity. EBE means Extra-Terrestrial Biological Entity.

2. **KRLL** was the first Alien Ambassador to the United States of America.

3. **GUESTS** were Aliens exchanged for Humans who gave us the balance of the YELLOW BOOK. At the onset in 1972 there were only three "aliens" left alive. It is estimated that there are presently 4,000 beings in the world who are referred as Alien Life Forms (ALFs) or OBS.

4. **YELLOW BOOK** is knowledge about Alien technology, culture, and their history.

5. **RELIGION**: The Aliens believe in a "Universal Cosmic God". They claim that "man" are Hybrids who were created by them. They claim all religion was created by them to hasten the formation of a "civilized culture" and to control the "human race". The Aliens have furnished proof of their claims and have a "device" that allows them to show audibly and visually any part of history that they or we wish to see.

6. **ALIEN BASES** exist in the four corner areas of Utah, Colorado, New Mexico and Arizona. Six bases were described in 1972, all on Indian Reservations, and all in the four-corner area. Dulce is one of the bases, and there are also bases in California, Nevada, Texas, Florida, Maine, Georgia, and Alaska.

7. **CRAFT RECOVERY** documents state that many craft had been recovered. The early ones came from Roswell, Aztec, Roswell again, Texas, Mexico and other un-named places.

8. **ABDUCTIONS** were occurring long before 1972, with early civilizations referring to these incidents. Documents state that humans and animals were being abducted and or mutilated, many of which vanished without a trace. Sperm, OVA, and tissue samples were taken, surgical operations were preformed, and implanted spherical devices were inserted into subjects 40 to 50 microns in size near the optic nerve in the brain. The document estimated that one in every 40 people had been implanted with this device.

The existence of Alien crafts at a hangar on Edwards Air Force Base was confirmed in 1989. The hangar is located at the north end of the base and is secured by guards who check the hangar each hour and report the status to NRO. The guards are instructed to never enter the hangar, even if it has been broken into. Authors of this document, "Blue Planet Project", have also confirmed the existence of "alien materials" at another special hangar at Edward's AFB.

CODE FACTS ABOUT ALIENS ON EARTH: OMNIDATA

The badge insignias that are on some Alien crafts and Alien Flags, is called

a **TRILATERAL** INSIGNIA (TRIADE).   **Found on some Spacecraft**

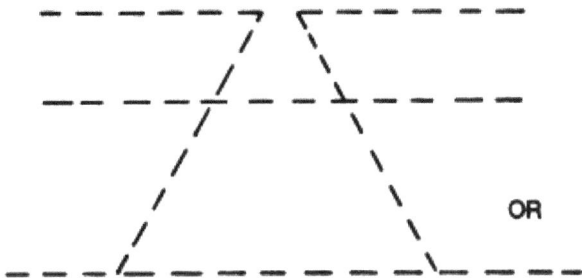

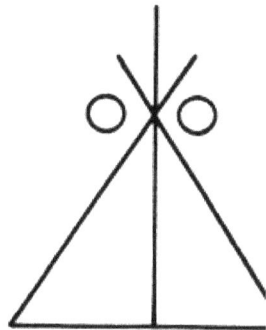

OR

Marks bases and Landing sites. This symbol is visible, "Only when viewed from Directly Overhead"

This symbol is found on some Regelians Spacecraft and their Uniforms

LUNA-2: Code name for the location of the second most important underground Rigelian base in New Mexico. The base is Alien controlled and NRO, DELTA, and Alien protected. Luna is ongoing.

FAR SIDE OF THE MOON: The term used in reference to inside LUNA-2. The LUNA-2 underground base is not a reference to the moon or a moon base.

"BLUE PLANET PROJECT": COMPILED MAJOR FACTS ABOUT ALIENS ON EARTH

- Alien crafts from other worlds have crashed on Earth.

- Alien crafts are from both Ultra-Dimensional sources and sources within this Dimension.

- Early U.S. Government efforts at acquiring technologies were successful.

- The U.S. Government has had live Alien hostages at some point in time.

- The U.S. Government has conducted autopsies on Alien cadavers.

- U.S. Intelligence Agencies and Security Agencies are involved in the cover-up of facts pertaining to the situation.

- People have been and are currently being abducted.

- There is a current Alien presence on this planet among us that controls different elements of our society.

- Alien forces maintain bases on Earth and on the Moon.

- The U.S. Government has had a working relationship with Alien Forces for some time, with the express purpose of gaining technology in gravitational propulsion, beam weaponry and mind control.

- Millions of cattle have been killed in the process of acquiring biological materials. Both Aliens and the U.S. Government are responsible for mutilations, but for different reasons.

- We live in a multi-dimensional world that is overlapped and visited by Alien entities from other dimensions. Many of those entities are hostile, and many are not hostile.

- The basis for our genetic development and religions lie in intervention by non-terrestrial and terrestrial forces.

- The truth about our actual technology far exceeds that perceived by the public.

- NASA and The United States Space program of today is a cover operation that exists for public relations purposes.

- People are being killed in order to suppress the facts.

- Our civilization is one of many that have existed in the last billion years, "Milky Way time period"

ALIEN PRESENCE ON EARTH

There are one hundred sixty or more known types of Aliens visiting Earth at the present time. The most commonly seen types:

- **Greys, type one**: The Rigelians from the Rigel Star system and are approximately four feet tall, with a large head containing large, slanted eyes, interested in technology.

- **Greys, type two**: Come from the Zeta Reticulae 1 & 2 solar systems. Same general appearance as a type one but have a different finger arrangement and a slightly different face. These Greys are more sophisticated than the type ones, possess a degree of common sense, and are somewhat passive.

- **Greys, type three**: Clones from types one and two, have thinner or no lips and are subservient to the type one and two Greys.

- **Nordics**: Are similar to humans, with blonde hair, blue eyes, will not break the law of non-interference to help us but intervene if the Greys activity were to affect us directly.

- **Nordic Clones:** They appear similar to us but with a grey tinge to their skin and are controlled drones, created by the Greys, type one.

- **Intra-Dimensional**: Entities that can assume a variety of shapes and are of a peaceful nature.

- **Short Humanoids**: One and a half to two and a half feet tall, with bluish skin, seen quite frequently in Mexico near Chihuahua.

- **Hairy Dwarfs**: Orange in color, red hair, four feet tall weighing about thirty-five pounds and seem to be neutral and respect intelligent life forms.

- **Tall Alien Race:** Resemble humans but are seven to eight feet tall, united with the Nordics.

- **Men in Black (MIBs)**: These aliens are oriental or olive-skinned, with eyes sensitive to light with vertical pupils. They wear black clothes, sunglasses and drive black cars. In groups they all dress alike. They very often intimidate UFO witnesses and impersonate government officials and are the equivalent of our CIA but from another Galaxy.

Editor's Note: To see a list and drawings of over 80 different Alien species read the book "The Extraterrestrial Species Almanac" by Craig Campobasso.

MAIN IMPORTANT UFO CRASHES UNDER ULTRA-SECRET COVER-UPS

- New Mexico: Roswell, San Augustine Flats, Aztec; 1947-1948: Affected by a new experimental radar type (JOSHUA and GABRIEL PROJECTS)

- California desert: 1952 cause unknown

- Mexico, Sonora Desert: 1955 cause unknown

- Florida, Everglades: 1959 cause unknown

- Texas, Dayton: 1961 cause unknown

- Brazil, Mato Grosso: 1966 cause unknown

- Uruguay, Salto: 1970 cause unknown

- Brazil, Amazon: 1976 cause unknown

- Mexico, Gulf of Mexico 1979 cause unknown

- Argentina, Andes: 1979 cause unknown

- Alaska, Fairbanks: 1981 UFO malfunction

"Blue Planet Project" disclaimer: When you read this document, it may disturb you. The real facts have a tendency to do exactly that, but this information needs to get to the general public. I caution you to use your best judgment when allowing a minor child to read this document.

Chapter 13

Deep Water Trenches

"Using increasingly sophisticated tools, technologies, and sensors, we learn more about our oceans every day; however, we have still explored less than five percent of this vast underwater frontier."
  -National Oceanic and Atmospheric
    Association

How deep is deep? It is almost unimaginable that we have absolutely no idea what lies deep underwater in our Oceans, Seas, and many lakes across the world. Scientists, the U.S. military and militaries of other nations, scuba divers, even the "National Oceanic and Atmospheric Association, are unaware of, an incapable of, reaching many sections of our Planet Earth. We know more about the moon than the deepest areas of our waterways. Ocean deep water trenches can be miles deep, far beyond the collapsible depth of our best nuclear submarines.

The Pacific Ocean

The Pacific Ocean is the largest and deepest ocean basin on Earth, covering more than 60 million square miles, averaging a depth of 13,000 feet and is 35,797 feet or 6.77 miles at its deepest. Covering more than 30 percent of the Earth's surface, the Pacific Ocean is the largest water mass on the planet. This makes for an ideal location for an underwater base, being human or alien. Protected by the depths of the Oceans and Seas, those with the intent to conceal themselves from satellite images,

aircraft and ocean bearing vessels need only submerge.

The Atlantic Ocean

The Atlantic Ocean is the second-largest ocean in the world, covering an area of approximately 41,100,000 sq miles. This ocean covers approximately 20 percent of the Earth's surface and approximately 29 percent of its water surface. The Atlantic Ocean has an elongated, S-shaped basin which extends longitudinally between Europe and Africa to the east, and North and South America to the west. It also extends north to the Arctic Ocean, to the Pacific Ocean in the southwest, the Indian Ocean in the southeast, the Southern Ocean in the south and southward to Antarctica. The northern Atlantic Ocean and the southern Atlantic Ocean is divided into two parts, by the Equatorial Counter Current, at about 8°N.

Deep Water Trenches

An oceanic trench is a long and narrow depression in the ocean floor. These trenches are considered the deepest part of the ocean floor, occurring at the boundary between convergent plates and lithospheric plates. These plates slowly move toward each other at distances that range from a few millimeters per year to more than 10 centimeters. The trench is created when one of the plates slides below the lithospheric slab. Generally, oceanic trenches reach between 1.9 and 2.5 miles below the nearby ocean floor. worldatlas.com

Figure 56 Debbie Ziegelmeyer diving in Cozumel 2006

In the depths of the oceans lie deep water trenches rarely visited and left untouched by man due to the deep-water pressures at depth. Very few attempts have been made to reach these dangerous depths. One recent explorer took the challenge on March 12, 2012. James Cameron, film director and explorer, hit bottom at 35,756 feet below the surface, staying a mere three hours before having to surface due to safety concerns. At depth, color disappears because water is 800 times denser than air causing it to absorb light extremely fast. The deeper you descend the duller color appears until finally, most color is absorbed. Red disappears first as shallow as 15 feet deep, followed by orange at 25 feet, yellow at 35 feet, and green at 70 feet. Blue is the last to go because it has the shortest wavelengths and highest energy. Pressure is a major factor at water depth for human as well as underwater vehicles such as submarines. Our atmosphere is 1 ATA and increases by 1 ATA for each additional 33 feet of descent. Basically, this mean that the deeper a diver descends the greater increase in air compression due to pressure, causing the air spaces in the diver's body to compress.

Pressure Chart
Surface: 1 ATA
33 feet deep: 2 ATA
66 feet deep: 3 ATA
99 feet deep: 4 ATA

On ascend, the reverse occurs decreasing pressure causing the air in a diver's air spaces to expand. By ascending slowly using required computed timed safety stops at various depths and amounts of time, a diver can safely return to the surface unharmed. Ascending too quickly not giving time for the diver's body air spaces to expand, could result in what is known has the "bends" or decompression sickness resulting in permanent injury or death.

Water at depth is difficult to transverse not only for a diver, but for a submarine. The Soviet submarine K-278 Komsomolets had a hull made of titanium and could descend as deep as 4265 feet but was very heavy and expensive to manufacture. The American's Seawolf class nuclear submarines are lighter weight high-grade steel and have an estimated crush depth of approximately 2400 feet.

What Happens to a Human Who Spends a Month Under the Sea?
By Brian Lam June 10, 2014, POPULAR SCIENCE Fabien Cousteau and his team are setting out to break the record for living in an underwater habitat. On June 1st, Fabien Cousteau, grandson of Jacques Cousteau, swam down to the last existing undersea habitat research lab in the world, Aquarius, in the Florida Keys. He'll live there for 31 days, which is a day longer than the time his grandfather's team spent living in his undersea habitat, Conshelf II, roughly 50 years ago. Since then, undersea bases have been created all over the world,

and have since lost their funding and ceased operation. Aquarius stands as the last.

The aquanauts joining Cousteau on "Mission 31" are photographers, scientists from Northeastern and MIT specializing in marine biology and underwater engineering, and Aquarius staff. Florida International University, the school responsible for saving Aquarius and operating it after NOAA cut funding last year, is of course sending down some of their own scientists, too. They'll all experience the unique challenge of living underwater for over a month in the pressurized saturated diving environment.

Saturated diving is a type of diving which allows the body to gradually soak up inert gases by staying at depth for a long period of time. These gases would harm a standard scuba diver by expanding like the bubbles in a shaken bottle of soda when the diver returns to the surface, causing pain, paralysis, and sometimes death. With the team sleeping in the base, at depth, and never surfacing, the divers are free to experience the most useful part of living in Aquarius: the ability to dive for 2-8 hours a day (as opposed to about an hour maximum per day that a regular scuba diver can achieve) without suffering from decompression sickness. At the end of the mission, the entire base is slowly brought back to normal pressure so that the gases can escape the diver's bodies safely, at which point the divers are free to resurface.

To find out what the human body and mind go through living in an underwater habitat, I spoke to various experts on living underwater, such as John Clark, Scientific Director for the U.S. Navy Experimental Diving Unit, who researches the effects of deep dives to 1,500 feet and Navy saturation diver Marc Chase who has worked on salvage jobs like the recovery of the USS Monitor's wreck. I also

spoke to Mark Patterson and Brian Helmuth, Mission 31 science advisors who have spent working time in Aquarius, and Mark Hulsbeck, the oceanographic field operations manager who will have spent 200 days in the base overall by the end of this mission. In the end, because research about long, relatively shallow underwater living is limited, there are a lot of theories as to the effects of living underwater on human beings, but much of it is controversial, anecdotal, and unproven by even those who study it and have experienced it. END

Water at depth is uninhabitable, and a very unfriendly region of Earth to mankind. This leaves a vast open area; 70 percent of the Earth with only 5 percent of which has been explored, available to those to use who have the means or more advanced technology. Those who transverse the heavens and universe with ease using technology possibly hundreds of years ahead of what is known to mankind, may not only be beneath the water's surface, but most likely have been for centuries. The ease in which these alien craft have been appearing suggests that they are not traversing space on a daily basis but are and have been residing on our planet for centuries. Alien races have found a new home far from their own home, which is private, safe, and for the taking.

Notable Trenches and Deepwater Areas

How many areas in our oceans and seas are left virtually unexplored, even today? Deep water trenches are located across the world and make perfect hiding places for alien bases, wormhole entries, vortexes, and time jumps. Some of the more notable trenches are listed here which not surprisingly, are active areas for UFO sighting reports.

Challenger Deep

The deepest location on the ocean floor is the Challenger Deep which reaches a depth between 35,755 and 36,197 feet below the surface. Located near the Mariana Islands of the Pacific Ocean within the Mariana Trench, The Challenger Deep has been descended on for exploration on four separate occasions. The first descent was by a manned bathyscaphe vehicle, the Trieste, and took place in 1960. The second descent was by an unmanned remotely operated underwater vehicle (ROV), named Kaiko in 1995, the ROV Nereus made the descent in 2009, and the most recent descent was in 2012 by the Deepsea Challenger, a manned deep submergence vehicle.

Mariana Trench

The Mariana Trench is the deepest point on the earth's surface, located in the western Pacific Ocean 125 miles east of the Mariana Islands, at a depth of 6.85 miles below sea level. This trench runs for 1,585 miles below the expanse of water between Australia and Japan. The majority of the Mariana Trench is now protected by the U.S. as part of the Marianas Trench Marine National Monument which was established by President George W. Bush in 2009. Permits for research in the area have to be obtained from the U.S. Fish and Wildlife Service. Because of its depth, the trench is extremely dark with the temperature just a few degrees above freezing. Water pressure at the bottom of the trench is a crushing eight tons per square inch which is about a thousand times the standard one atmospheric pressure at sea level.

On March 12, 2012, James Cameron, film director and explorer with the support of his crew, became the first human to travel on his own to the world's deepest abyss. Cameron descending in The Mariana Trench, took the Challenger to a depth of 35,756 feet deep which was confirmed by members of the National Geographic Expedition. The dive had been attempted only once before, in 1960, when Don Walsh, a retired United States Navy captain, and Jacques Piccard, a Swiss engineer, reached the same depth in the Navy submersible Trieste. Cameron's sub, the Deepsea Challenger, hit bottom at 35,756 feet below the surface where the water pressure is around 16,000 pounds per square inch, nearly the atmospheric pressure on the surface of Venus. He remained for a mere three hours of the nine he had planned due to safety concerns.

Tonga Trench

The Tonga Trench is the second deepest trench in the world lying about 6.76 miles below sea level. It is 850 miles in length, lying parallel with the eastern shore of Australia and can be joined with islands of New Zealand with a straight line on a map. This trench is located in the southwestern Pacific Ocean in the southern hemisphere and like the Mariana Trench, has been explored but only limited.

Galathea Depth

The third deepest site in the world is The Galathea Depth. Located in the western Pacific Ocean, it is within the Philippine Trench. The trench is a deep 34,580 feet below sea level on the ocean floor. In 1950, a Danish exploration team, whose original objective was to collect fauna from the bottom of the ocean, were among the first explorations of this site.

Kuril- Kamchatka Trench

The Kuril- Kamchatka Trench, also in the Pacific Ocean, lies at a depth of 6.5 miles below sea level, close to Kuril Island and off the coast of Kamchatka. This trench is very active with ocean bed volcanic activities.

Philippine Trench

The Philippine Trench is the fourth deepest part of the ocean in the world. It also lies in the Pacific Ocean at a depth of about 6.5 miles just east of the Philippines in the Philippine Sea at a depth of 34,580 feet. Located in the western North Pacific Ocean, it continues NNW-SSE, and stretches an amazing 820 miles. The trench's width is approximately 19 miles from the center of the Philippine Island of Luzon, going southeast to the northern Maluka island of Halmahera in Indonesia. Ten 7.2 magnitude earthquakes have occurred in this trench in the last 100 years.

Kermadec Trench

The fifth deepest part of the Pacific Ocean is the Kermadec Trench at a depth of 6.25 miles stretching to about 750 miles in length and forming the eastern boundary of the Kermadec Ridge. It's located in the south Pacific Ocean just above the northern coastline of New Zealand.

The Izu-Ogasawara Trench

The Izu-Ogasawara Trench also known as Izu-Bonin Trench, is 32,087 ft at its deepest, and located in the western Pacific Ocean. It extends from Japan to the northernmost section of the Mariana Trench and is an extension of the Japan Trench. In this underwater location, the Pacific Plate is being subducted beneath the Philippine Sea Plate, creating the Izu Islands and Bonin Islands on the Izu-Bonin-Mariana Arc system.

The Japan Trench

The Japan Trench is part of the Pacific Ring of Fire off northeast Japan, extending 497 miles from the Kuril Islands to the northern end of the Izu Islands. At 26,398 ft at its deepest, it links the Kuril-Kamchatka Trench to the north and the Izu-Ogasawara Trench to its south. The Japan Trench is one of the primary concerns of tsunamis and earthquakes in

northern Japan. On March 11, 2011, the Tohoku earthquake, Japan's strongest earthquake in its recorded history, resulted in a large tsunami causing mass destruction. The earthquake struck below the North Pacific Ocean, 81 miles east of Sendai, the largest city in the Tohoku region which is a northern part of the island of Honshu. The Tohoku tsunami produced waves up to 132 feet high, killing more than 15,500 people and rendering more than 450,000 people homeless. Included in the mass destruction was damage to and the meltdown of three nuclear reactors at the Fukushima Daiichi Nuclear Power plant causing the release of toxic, radioactive materials into the environment and causing the evacuation of thousands of people.

Puerto Rico Trench

The Atlantic Ocean is home to the Puerto Rico Trench which is the deepest part of the Atlantic Ocean and sixth deepest underwater point in the world. This trench is the boundary between the Caribbean Sea and the Atlantic Ocean. At a depth of 5.2 miles with a length of over 497 miles, this trench has been responsible for tragic tsunami and earthquake activity in this region. In 1953 an Earthquake of 8.1 on the Richter scale was registered which was one of the highest values ever recorded in the region.

South Sandwich Trench

The second deepest point found in the Atlantic Ocean is the South Sandwich Trench which lies 62 miles to the east of the South Sandwich Islands. With a depth of approximately 5.2 miles and stretching for over 594 miles, The South Sandwich Trench it one of the most noticeable trenches of the world, V-shaped, and is associated with an active volcanic arc.

Romanche Trench

The Romanche Trench, lying near the equator in the mid-Atlantic Ocean, is the third deepest trench of the Atlantic. This trench is in the list of top fifteen deepest parts of the oceans of the world lying at a depth of 25,453 feet running east-west between the shoulders of South America and Africa. It stretches 48.7 miles in length and plays a major role in uniform distribution of water in this particular part of Atlantic Ocean.

Java Trench

The Java Trench at a depth of 4.8 miles, is the deepest part of Indian Ocean. Also referred to as Sunda trench, it lies close to the islands of the Indian Ocean including Java, Sumatra, and Andaman Islands. It extends 2,000 miles in a northwest south-west arc along the southwestern and southern Indonesian archipelago.

Eurasian Basin

Found under the frozen waters of Arctic Ocean, The Eurasian Basin is one of the two major basins into which the North Pole Basin of the Artic Ocean is split by the Lomonosov Ridge. Lying at a depth of 3.4 miles and extending 217 miles, the basin is bounded from the south by Greenland and the Svalbard archipelago, the Lomonosov Ridge, and the Siberian shelves of Laptev Sea, Kara Sea and Barents Sea. This trench is the deepest point of the Arctic Ocean and the twentieth deepest part of any ocean in the world.

Amerasian Basin

The Amerasian Basin is the other basin besides the Eurasian, in which the North Polar Basin of the Artic Ocean is split by Lomonosov. Its triangular shape extends from the Canadian Artic Islands to the East Siberian Sea and from Alaska to the Lomonosov Ridge.

Chapter 14

Why Are Alien Races Colonizing Earth's Waterways?

Undetectable Travel

Seventy percent of the Earth is comprised of water which is spread across the entire planet. In some areas, one could travel undetected conversing the entire planet without ever breaking surface. In the state of Missouri where I reside, rivers, lakes, creek, and streams are so widespread that if mapped out precisely, underwater travel to almost anywhere in the state could be accomplished.

Underwater military travel in wartime began in 1620 but was more commonplace and submarine-like in the late 1800's. Unobstructed underwater travel is possible today by nuclear class submarines from every world power nation's Navy. They can transverse every ocean and sea undetected remaining submerged indefinitely. Colombian drug traffickers are also using submarines as a new secret means of transporting drugs north. Cocaine is being smuggled in submarines up to 100 feet long which are nearly impossible to detect. These submarines are capable of distributing several tons of cocaine in just one shipment. Dozens of submarines are thought to be in operation between the coasts of Colombia and Mexico, with law enforcement estimating that another seventy will be built in the next year alone. There are also at least 300 known cocaine finishing labs and numerous drug submarines with manufacturing factories.

Reports from submarines of "fast movers" have been hidden from the public for decades. My friend Marc D'Antonio, has been involved with MUFON since 1971 and has a vast amount of experience investigating UFO cases. He is also the owner of FX Models where work is done on CGI and physical models for the entertainment industry and defense contractors. In 1994, Marc D'Antonio was working on a classified government project which involved going to sea on a U.S. nuclear submarine. While at sea Marc became seasick. Apparently when seasick, one if permitted, moves to the sonar room where there is less at sea movement. While there, the sonar operator picked up a fast-moving object on his screen. The sonar operator informed the XO (executive officer) about the sighting and asked what he should do. The answer was to log it and store it. Marc being the proclaimed "UFO expert" on the ship asked the XO if he could assist. The reply was "enjoying your time on my boat"? Marc replied "yes". The XO followed with the reply, "Want to stay?" Marc realized from the reaction that the matter was closed, but Marc also knew he had seen what is known as a "fast mover." A few years later Marc met an Admiral and his wife at a dinner.

Marc, being the ever-inquisitive investigator he is, didn't miss the opportunity to ask the Admiral about what he had seen in the sonar room while on that classified mission. Marc asked, "Admiral, what can you tell me about the Fast Mover Project?" Marc stated that the Admiral smiled and replied "Marc, you know I can't talk about that." Since being told the prior story by Marc D'Antonio, I have taken every opportunity to ask those I meet who have served aboard submarines if they had seen or experienced anything similar. To date, I have seen many reports, mostly Russian, and have received two "yes" replies.

Claims regarding Russian and U.S. Navy nuclear submarines detecting and even interacting with UFOs or USOs, Unidentified Submerged Objects, have been rumored for decades. These underwater vessels are equipped with some of the most sensitive listening devises on the planet, such as sonar arrays and computer systems in the hundreds of millions of dollars. Tom DeLonge, To the Stars Academy, released a statement claiming that a few years ago an unidentified craft was underwater and pinned against the North Atlantic coast by multiple nuclear attack submarines for over a week. According to released top secret documents, Russian submarines are fighting a secret war with "alien" craft deep under the oceans. According to a book titled "Russia's USO Secrets", by investigator Philip Mantle which is based on Russian documents and accounts from military veterans, Russians are playing a game of cat and mouse with strange underwater craft.

One event describing an incident in the Bermuda Triangle in 2009, captained by former nuclear submarine commander Yury Beketov who stated, "We repeatedly observed that the instruments detected the movements of material objects at unimaginable speed, around 230 knots (400 km per hour, or 250 mph). It's hard to reach that speed on the surface, only in the air is it possible. The beings that created those material objects significantly exceed us in development." Lieutenant-Commander Oleg Sokolov revealed to his students that while on duty, he spotted a strange object rising out of the water through the submarine's periscope.

Marine scientists in the port city of Sevastopol, Ukraine, reported sighting a huge "wheel" rotating below the Black Sea while they were deep sea diving.

Documented in records in 1951, a Soviet submarine encountered a gigantic underwater object heading towards the shores. The captain ordered depth bombs to be dropped into the path of the unknown vessel, but the vessel did not react to the attack and stayed on its course before leaving at a high rate of speed towards the surface. When at a depth of 50 meters, it stopped its ascent, changed course, and departed off into the distance.

Released Russian files reveal that In July of 1978, Russian sailors observed an unknown vessel in the Mediterranean Sea. The captain of Soviet ship The Yargora, immediately sent a radiogram to the Soviet Academy of Sciences in Moscow informing them that the unknown vessel was shaped like a flattened-out sphere and was a white pearl color.

Hydrogen

Hydrogen is high in energy, yet an engine that burns pure hydrogen produces almost no pollution. NASA has used liquid hydrogen since the 1970s to propel the space shuttles and other rockets into orbit. Hydrogen fuel cells power the shuttle's electrical systems, producing a clean byproduct - pure water, which the crew drinks. A fuel cell combines hydrogen and oxygen to produce electricity, heat, and water. Fuel cells are often compared to batteries. Both convert the energy produced by a chemical reaction into usable electric power. However, the fuel cell will produce electricity as long as fuel (hydrogen) is supplied, never losing its charge.

Hydrogen Fuel Cells

NASA has for decades, relied upon hydrogen gas as rocket fuel to deliver crew and cargo into space. By means of the Centaur, Apollo and space shuttle vehicles, NASA has developed extensive experience in safe and effective handling of hydrogen. Hydrogen, in the use of the rocket engines on each shuttle flight burn, burn about 500,000 gallons of cold liquid hydrogen. Another 239,000 gallons are depleted by storage boil off and transfer operations. With missions to the moon and eventually Mars in the planning stages, hydrogen will continue to be innovatively stored, measured, processed, and employed. Hydrogen can be derived from local water or soil to supply fuel for transportation, electrical power, and crewmember breathable oxygen. This experience with fuel cell power and life support systems, provides an additional basis for future exploration. On the International Space Station, water is split into oxygen for breathing and hydrogen for fuel, and in the future, hydrogen will be recombined with exhaled carbon dioxide for water renewal. By generating and recycling hydrogen in space, cost factors will decrease the cost and missions to the ISS will be reduced because the need for supplies delivered from Earth will lesson. In recent years, the Department of Energy (DOE) and private industry have made significant advances in the development of Proton Exchange Membrane (PEM) fuel cells using hydrogen and air as the fuel and oxidant for ground-transportation applications.

Carl W. Feindt: "UFOs and Water"
Physical Influences of A UFO On Water

"The surface of any body of water is the physical boundary between air and water, a boundary not yet mastered by modern technology. Water depth involves extreme pressure.

1. Is it possible that underwater UFOs while moving submerged, never come into contact with the water surrounding them because of a surrounding magnetic field?
2. Speculation: In flight, the surrounding air moves along with the UFO accounting for the absence of sonic booms at high speed. It is possible that water behaves similarly, and this lack of friction explains why the craft can perform as well or better underwater than a submarine?
3. Is the UFOs powerful magnetic propulsion field polarizing the water vapor on the area's surface, causing it to form a fog or mist?"

Alien Use of Hydrogen

Is it possible that alien craft are using the Earth's water to harvest Hydrogen as a mean of replenishing some type of fuel cell? Water, H_2O, is composed of two atoms of hydrogen and one atom of oxygen, making water properties unique and essential to life on Earth. Oxygen, naturally, is also essential to human life but in a different manner. Carl Feindt spoke of the physical influences of a UFO on water pointing out that "alien" craft never actually come in contact with water due to their electromagnetic field. But what is generating that field? There have been sightings of "unknown vessels" logged by seamen, which are hovering over, going into, and coming out of large bodies of water for centuries, some doing so at a high rate of speed. Even with modern day technology, aircraft traversing from air to water even at a much lower rate of speed has a disastrous outcome. Sonar reports from submarines of "fast movers" are proof that unidentified craft with technology far surpassing our own, not only travel underwater but do so at a fast rate of speed with ease.

Harvested Food Source

Over the years there have been several reports of alien craft hovering over water sources or seen just below the water surface, sucking up the water from below then expelling part of the accumulated water from the craft. This could be a means of collecting Hydrogen, the process of a cooling type system, or the process of collecting some type of food source from the water as a means of nourishment, or both? Although unverified, alien species are thought to be primarily vegetarian.

Plankton are marine drifters better known as organisms, which are carried along by tides and currents. It is the productive base of both marine and freshwater ecosystems, providing food for smaller fish and sea mammals such as dolphin and whales. Although they are microscopic in size, they provide the base for the entire marine food web. Unlike fish, plankton do not swim on their own, they drift with tide and current and are present in some type of form in both fresh and saltwater areas.

Seaweed from oceans and seas can be consumed as food and are available in three different types, brown, red, and green. Other aquatic plants available in both salt and sea water include watercress, water chestnuts, and various other edible plant species.

Ocean "phosphorescence" is commonly seen at night when the water is disturbed largely due to the dinoflagellates which occur in the oceans as planktonic forms. These respond to stimulation when the water is disturbed by emitting brief bright light. Light emission may

be seen in the wake of a large ship for some 20 miles and is a favorite experience for night scuba divers as they surface. Approximately twenty percent of marine species are bioluminescent, and many are photosynthetic. "Red tides" which are transient blooms of individual dinoflagellate species, are often seen during daylight hours. These are found on the Pacific coast and in the Caribbean. These may also be a sought-after food source for alien life.

Chapter 15

RELEASED UNDER FOIA
(Freedom of Information Act)

AFOSR 68-1656 UFOs and Related Subjects
This document has been approved for public release and sale; Its distribution is unlimited.
Library of Congress Card Catalog No. 68-62196

UFOs and Related Subjects:

An Annotated Bibliography Lynn E. Catoe Prepared by the Library of Congress Science and Technology Division for the Air Force Office of Scientific Research Office of Aerospace Research, USAF Arlington, Virginia 22209 under AFOSR project orders 67-0002 and 68-0003

U.S. Government Printing Office, Washington: 1969
For Sale by the Superintendent of Documents U. S. Government Printing Office Washington, D. C. 20402
Price $3. 50

FOREWORD
The subject of unidentified flying objects is a popular phenomenon of the period from 1947, evoking widespread speculation and producing a literature of great variety and scope. This body of literature and documentation is source material for readers seeking better understanding of this question, which has involved the U.S. Air Force as well as other official groups. This is believed to be the most comprehensive bibliography published to date on the subject and includes the extensive UFO collection of the Library of Congress, as well as related material useful in understanding the nature of the question. The bibliography was produced by the Library's Division of Science and Technology with support provided by the Air Force Office of Scientific Research, a unit of the Office of Aerospace Research, the research agency of the U.S. Air Force. The bibliographer is Miss Lynn E. Catoe, who also collected the books, journal articles, pamphlets, conference proceedings, tapes, original manuscripts, and other material listed here, a total of more than 1,600 items. This literature survey was requested by AFOSR to assist a scientific research project at the University of Colorado under the direction of Dr. Edward U. Condon on V unidentified flying objects. The research began 1 November 1966 and has been carried out with support provided by AFOSR at the direction of the Secretary of the Air Force under contract F44620-67-C-0035. The preparation of the bibliography was accomplished under AFOSR project orders 67-0002 and 68-0003.

Unidentified Submarine Objects

Binder, Otto O. The mystery of flying saucers at sea. Rudder, v. 84, Feb. 1968: 21-23, 75. Account of UFOs observed at sea and suggestions of how boatmen can help solve the phenomenon.

Bowen, Charles. A South American trio. Flying saucer review, v. 11, Jan.-Feb. 1965: 19-21.
   Three UFO sighting cases from South America:
   (1) In April 1957, a resident of Cordoba, Argentina, allegedly encountered a landed flying saucer and was invited by

one of the crew members to enter for an inspection tour.

(2) On Jan. 10, 1958, an unidentified floating object was viewed off the coast of Paulo, Brazil by several witnesses before sinking out of sight.

(3) During a fire near Sao Bernado do Campo, Brazil, in 1963, a flying saucer landed amid the flames, and several tall, good-looking "people" emerged from it and picked up pieces of burnt material, stones, and other debris.

Fouere, Rene. Existe-t-il des bases sous-marines de soucoupes volantes? Phenomenes spatiaux, Feb. 1965: 16-25. Lists instances 1345-1960 where disc-like or wheel-like glowing objects were seen entering the oceans, in or on the oceans, or leaving the oceans. Gives reasons the ocean depths would be ideal flying saucer bases.

Galindez, Oscar A. Crew of Argentine ship see submarine UFO. Flying saucer review, v. 14, Mar.-Apr. 1968: 22

A "submergible UFO with its own illumination" allegedly paced the Argentine steamer, Naviero, for 15 minutes on July 30, 1967. Article taken from press reports in the Argentine newspapers La Razon, Cordoba, and Los Principios.

Hinfelaar, H. J. Submarine craft in Australian waters. Hying saucer review, v. 12, July-Aug. 1966: 28-30.

Reports appearances of Unidentified Submarine Objects (USOs) in New Zealand and Australian waters in 1965.

Ley, Willy. The wheels of Poseidon; and too much imagination. In For your informal; one on-earth and in

the sky. New York, Doubleday & Co., 1967. p. 69-88, 157-168.

A phenomenon is described that occurs in the Indian Ocean and that has been described by several eyewitnesses: a pulsating near-circular disturbance roughly 1,000-1, 500 feet in diameter is seen with streaks of light like the beams of a searchlight radiating from its center and revolving counterclockwise. Views are given supporting speculation that the Fodkamennaya Tiinguska Meteorite of Central Siberia (190b) was in fact an exploded extraterrestrial spacecraft.

Lorenzen, Coral. Diving for lost UFO. Fate, v. 17, May 1964: 62-65

Efforts are being made to salvage a flying saucer which seemed in mechanical difficulty when it allegedly sank in the Peropava River, Brazil, on October 31, 1963.

Luminous wheels puzzle seamen. New scientist, Mar. 9, 1967, v. 33:447-448.

In March 1966, three merchant ships in the Gulf of Thailand independently observed the apparently unexplained phenomenon known as the phosphorescent wheel: bands of luminosity skimming across the surface, apparently radiating from a central bright source. The wheels can rotate in either direction, and there have been reports of two wheels, one above the other, rotating in opposite directions. Professor Kurt Kalie of Hamburg, authority on phosphorescent wheels, attributes the phenomenon to bioluminescence.

Luminous wheels puzzle seamen, Spacelink, v. 4, Summer 1967: 12-13. In the Gulf of Thailand and waters to the south-east during March 1967, three merchant ships independently observed the unexplained phenomenon known as the phosphorescent wheel: bands of luminosity apparently radiating from a central bright source. Professor Kurt Kalie of Hamburg, authority on luminescent wheels, attributes them to bioluminescence. From New Scientist, March 9, 1967.

Ribera, Antonio. More about UFOs and the sea. Flying saucer review, v. 11, Nov.-Dec. 1965: 17-18.

Author summarizes events which may indicate underwater reconnaissance by UFOs of submarine bases.

Ribera. Antonio. UFOs and the sea. Flying saucer review» v. 10, Nov.-Dec. 1964: 8-10.

Account of strange happenings at sea that might support hypothesis that the bodies of water covering three-quarters of earth's surface are providing a hiding place for UFOs.

Robertson, W. S. UFOs, and the Scottish seas. Flying saucer review, v. U, May-June 1965: 36-37,

Account from newspaper sources of UFO sightings in the waters off the coast of Scotland, 1961-1965.

Steiger, Brad and Joan Whritenour. Unidentified underwater saucers. Saga v 36 June 1968: 34-37, 54-57.

Instances are cited in which UFO seen hovering over oceans, lakes, and rivers have submerged in the water. It is suggested that the objects may have underwater "bases,"

Turner, Richard. Some unfamiliar 'PSUFOs': the phosphorescent wheels. Flying saucer review, v. 13, Sept.-Oct. 1967: 7-9.

Account of unusual bioluminescent phenomena that may be mistaken for UFOs. UFOs sightings by seamen (1875-1910) of "phosphorescent wheels" that may fall into this category.

Released: F.O.I.A.

ANTARCTIC FLYING SAUCERS: A
group of red, green, and yellow flying saucers has been seen flying over Decepcion Island for two hours by Argentine, Chilean, and British bases in Antarctica. The flying saucers were also seen flying in formation over South Orkney Islands in quick circles. (Buenos Aires ANSA Spanish 1556 ORT 6 July 1965—P)

September 23, 1951: UFOs Over
Long Beach

Documents released under FOIA:

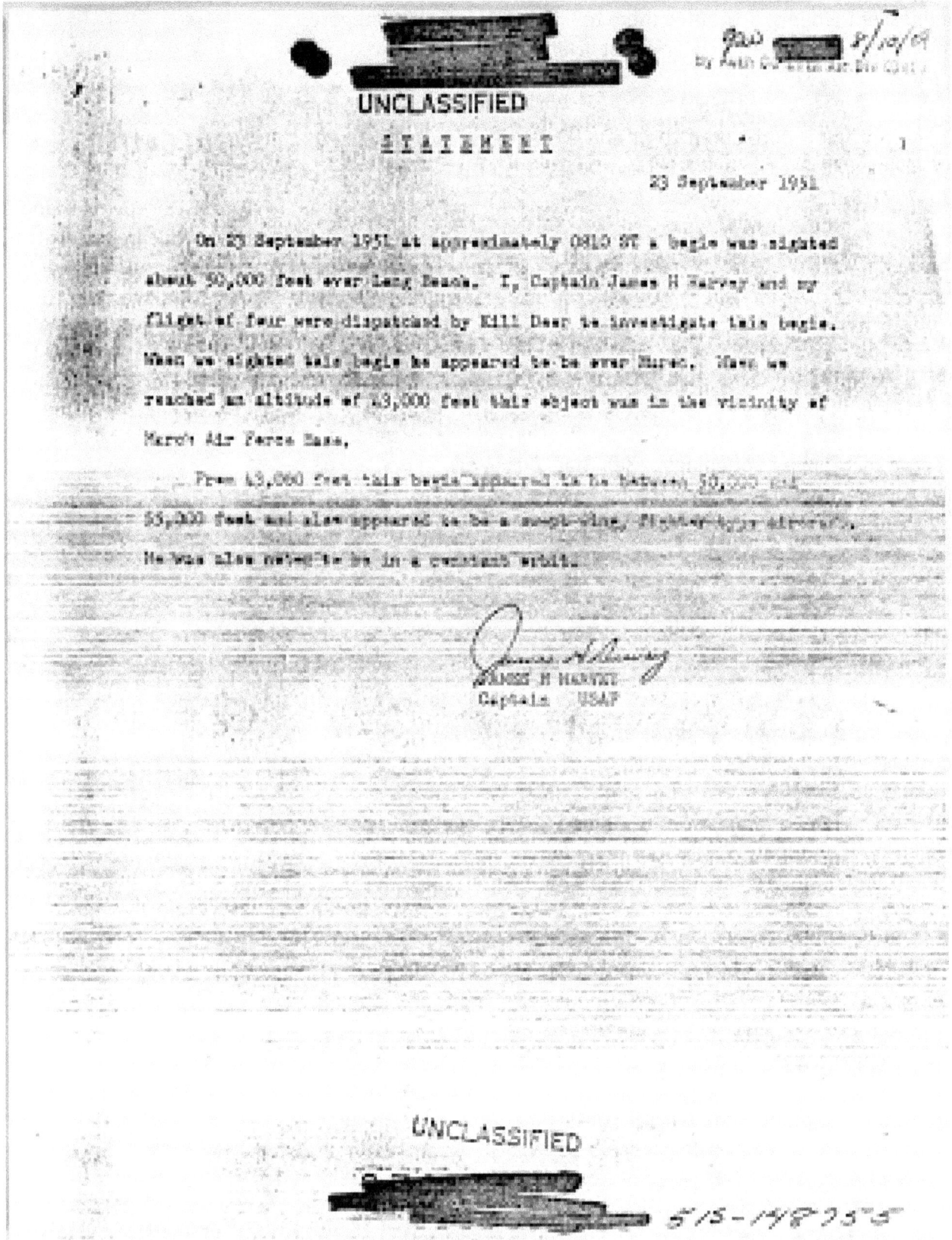

UNCLASSIFIED

S T A T E M E N T

23 September 1951

On 23 September 1951 at approximately 0810 ST a bogie was sighted about 50,000 feet over Long Beach. I, Captain James H Harvey and my flight of four were dispatched by Kill Deer to investigate this bogie. When we sighted this bogie he appeared to be over Hurac. When we reached an altitude of 43,000 feet this object was in the vicinity of March Air Force Base.

From 43,000 feet this bogie appeared to be between 50,000 and 55,000 feet and also appeared to be a swept-wing, fighter type aircraft. He was also noted to be in a constant orbit.

JAMES H HARVEY
Captain USAF

UNCLASSIFIED

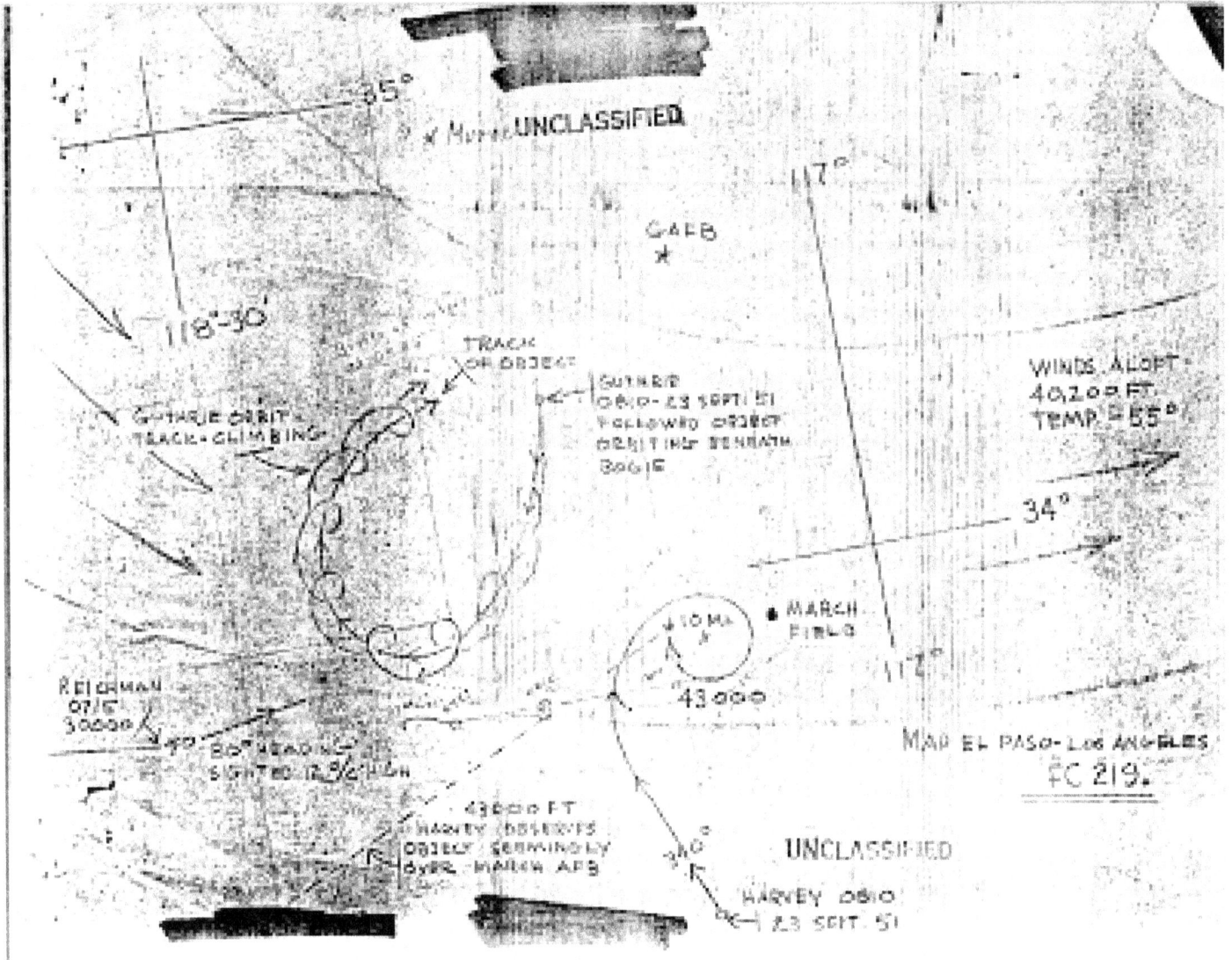

Figure 57 Page 2 UFOs over Long Beach

Project Sun Streak

In the 1970's, governments including the United States, were experimenting with the use of remote viewing as an intelligence tool. Experiments were conducted using remote viewing as a means of intelligence gathering from inaccessible areas across the world. The project continued into the 1980's, with the CIA taking charge forming Project Sun Streak.

Project Sun Streak was establish describing it as a "DoD-approved special access program." Eleven people were assigned to the project to "pursue the practical applications of applied psychoenergetics against selected operational tasks and improved skills through proven training techniques." The team goal was to be given tasks in hopes of "cueing Intelligence Community Sources in the collection of foreign intelligence." Project Sun Streak Senior executives were comprised of a panel which included the DIA, CIA, AIA, AF/In, DNI, and C3I, who oversaw, monitored, and advised on matters of concern and projected targets.

Project Sun Streak members were given a description of an area or specific location and asked to provide details on activity or descriptions of the inside of structures. Many of the documents I found seemed to describe Russian activity and bases in the Arctic. Others focused on areas in the middle east including troop movement, bases, and even a battle. However, some of the documents leaned towards inquiries into unknown, unidentified activity and a possible area of interest located in a mountainous area.

Since Alien base locations have been suspected to be located inside remote, hard to access, mountainous areas, I felt a couple of the documents in this chapter are of interest and warrant review. I also find it of great interest that a country such as the United States who is said to have one of the most powerful military forces in the world, is using remote viewing as a tool of intelligence gathering. Is it possible that the remote viewing of a Russian base or troops in the middle east are actually training sessions for intelligence gathering of something much more unknown and inaccessible? Could the actual goal of the CIA be in the pursuit of intelligence on alien bases?

The Black Vault

Thank you to the John Greenewald, Jr. for providing the following documents he obtained under the Freedom of Information Act for The Black Vault, and for allowing me to reprint them here:

CIA Documents:

Figure 59 •   C00015435 February 23, 1967 Memorandum: $300,000 Condon contract

C00015237

C̶O̶N̶F̶I̶D̶E̶N̶T̶I̶A̶L̶ 141445Z APR 76 STAFF

CITE DCD/▮▮▮▮▮

TO: PRIORITY DCD/HEADQUARTERS.

ATTN: ▮▮▮▮▮

FROM: DCD/▮▮▮▮▮

SUBJECT: CASE ▮▮▮▮ - UFO RESEARCH

REF (A): DCD/HEADQUARTERS 14596

(?): FORM 618 DATED 9 APRIL 1976, UFO STUDY.

1. SOURCE'S FULL NAME IS ▮▮▮▮▮
HE IS EMPLOYED AS ▮▮▮▮▮

2. REFERENT B MATERIAL CLASSIFIED CONFIDENTIAL AT HIS
REQUEST. SOURCE SEEKS GUIDANCE FROM CIA UFO EXPERTS AS TO
MATERIAL IN HIS REPORT THAT SHOULD REMAIN CLASSIFIED.

APPROVED FOR RELEASE
DATE 17 nov 78

Figure 58 •1976 UFO Study Memo C00015237

C00015339

OCT 2 1952

MEMORANDUM TO: Director of Central Intelligence

THROUGH: Deputy Director (Intelligence)

FROM: Assistant Director, Office of Scientific
                    Intelligence

SUBJECT: Flying Saucers

1. PROBLEM—To determine: (a) Whether or not there are national
    security implications in the problem of "unidentified
    flying objects"; (b) whether or not adequate study and
    research is currently being directed to this problem
    in its relation to such national security implications;
    and (c) what further investigation and research should
    be instituted, by whom, and under what aegis.

2. FACTS AND DISCUSSION—OSI has investigated the work currently
    being performed on "flying saucers" and found that the
    Air Technical Intelligence Center, DI, USAF, Wright-
    Patterson Air Force Base, is the only group devoting
    appreciable effort and study to this subject, that ATIC
    is concentrating on a case-by-case explanation of each
    report, and that this effort is not adequate to corre-
    late, evaluate, and resolve the situation on an over-
    all basis. The current problem is discussed in detail
    in TAB A.

3. CONCLUSIONS—"Flying saucers" pose two elements of danger
    which have national security implications. The first
    involves mass psychological considerations and the
    second concerns the vulnerability of the United States
    to air attack. Both factors are amplified in TAB A.

4. ACTION RECOMMENDED—(a) That the Director of Central Intel-
    ligence advise the National Security Council of the
    implications of the "flying saucer" problem and request
    that research be initiated. TAB B is a draft memo-
    randum to the NSC, for the DCI's signature. (b) That
    the DCI discuss this subject with the Psychological
    Strategy Board. A memorandum to the Director,
    Psychological Strategy Board, is attached for sig-
    nature as TAB C. (c) That CIA, with the cooperation
    of PSB and other interested departments and agencies,
    develop and recommend for adoption by the NSC a

25

Figure 60 •C00015339 Memorandum: Flying Saucers

policy of public information which will minimize concern
and possible panic resulting from the numerous sightings
of unidentified objects.

H. MARSHALL CHADWELL
Assistant Director
Scientific Intelligence

ANNEXES:
    TAB A—Memorandum to DCI, through DDI, Subject: Flying
        Saucers.
    TAB B—Letter to National Security Council with enclosure.
    TAB C—Memo to Director, Psychological Strategy Board with
        enclosure.

CONCURRENCES:

Date: _____

LOFTUS E. BECKER
Deputy Director/Intelligence

ACTION BY APPROVING AUTHORITY:

        Date: _____

Approved (disapproved): _____

WALTER B. SMITH
Director

- 2 -

C00207070

UNCLAS /TAB

SOV ECON POLICIES, PROBLEMS (4 MIN: MAND 0900)
    1518.   STATION REPORT ON RESTORATION OF DAMAGED CATHEDRALIN
MOSCOW (6 MIN: MAND 0900)
    1519.   USSR TODAY: REPORT ON SOV ISLAND IN ARTIC CIRCLE WHICH
ONCE HAD MANY CONCENTRATION CAMPS (4 MIN); REPORT OF SIGHTING OF UFO
IN KRASNOYARSK (4 MIN) (TOTAL 8 MIN: MAND 1300)
    1520.   WANG XIAO HALF-HOUR PROGRAM: HIGHLIGHTS OF SOV WEEKLIES
ARTICLES (3 MIN); INTVW WITH CHMN OF PRIVATE BUSINESS ASSOC (4.5
MIN) (TOTAL 7.5 MIN: MAND 1300)
    1521.   PRAVDA'S VIEWS LATEST USSR SUPSOV SESSION (1 MIN: MAND
1400)
    1522.   REPORT ON SUPSOV DISCUSSIONS ON UNION TREATY (5.5 MIN: JAP
141400)
    1523.   MISC INTERNAL USSR ITEMS: JAP 2-141100 141400
    UNPRO NONE: POOREST NONE
(ENDALL) ~~████████████~~ 121607.008
BT
#3457
NNNN
NNN

UNCLAS /TAB

Figure 61 •C00207070 Arctic Circle UFO sighting

C00179737

A RUSSIAN OFFICER. IT IS A PROVOCATIVE QUESTION. BUT SUCH ACTIONS ARE NOT ONLY BEING ADDRESSED TO SPECIFIC PERSONS. THE ORGAN OF THE UKRAINIAN NATIONAL ASSEMBLY, UKRAINSKI OBRII, WRITES AS FOLLOWS ABOUT THE ROLE OF THE ANDREYEV FLAG, WHICH IS SACRED FOR THE RUSSIAN SAILOR: "FOR 200 YEARS THE ANDREYEV FLAG COVERED UP THE ROTTENNESS OF THE RUSSIAN EMPIRE. NOW THEY WANT IT TO COVER UP THE CRETINISM OF THE CIS. ENOUGH!"

BUT, GENERALLY SPEAKING, I DO NOT BELIEVE THAT THE UKRAINIAN PEOPLE ARE CAPABLE OF ADOPTING A HOSTILE ATTITUDE TOWARD RUSSIA. WE MAY SPEAK, THEREFORE, OF A WHOLLY NEGLIGIBLE, RECKLESS GROUP OF UKRAINIAN NATIONALISTS, CHIEFLY OF A WESTERN UKRAINIAN PERSUASION. AND UNLESS THERE IS A CHANGE IN THE VERY NEAR FUTURE IN THE PRESENT POLICY COURSE, I BELIEVE THAT BIG INTERNAL UPHEAVALS AWAIT UKRAINE.

((UGLANOV)) ARE CONSCRIPTS BEING DRAFTED FOR RUSSIA FROM OTHER REPUBLICS?

((KOTENKOV)) NOT AS YET. ALTHOUGH AT THE RECENT COUNCIL OF DEFENSE MINISTERS ON 28-29 APRIL, WHICH WAS HELD IN MOSCOW, IT WAS A QUESTION OF A NUMBER OF CIS STATES BEING PREPARED TO MAKE THEIR CONSCRIPTS AVAILABLE FOR SERVICE IN THE CIS JOINT ARMED FORCES, ON THE TERRITORY OF RUSSIA INCLUDED.

((UGLANOV)) WILL COSSACK UNITS BE FORMED IN THE SPHERE OF THE DON ARMY?

((KOTENKOV)) THERE IS NO FULLY FINISHED CONCEPT OF THE ORGANIZATIONAL DEVELOPMENT OF THE RUSSIAN ARMED FORCES AS YET. A STRUGGLE OF IDEAS IS UNDER WAY, AND I WOULD NOT CARE TO DWELL ON THIS. BUT SUCH AN ASPECT OF MILITARY ORGANIZATIONAL DEVELOPMENT IS POSSIBLE. IT COULD BE A QUESTION OF PROFESSIONAL RESERVES. WITHIN THE FRAMEWORK OF SUCH A PROFESSIONAL RESERVE I WHOLLY ALLOW OF THE POSSIBILITY OF THE EXISTENCE OF COSSACK UNITS. WITH THE OBSERVANCE OF COSSACK TRADITIONS, SOLE RESPONSIBILITY, FIRM DISCIPLINE AND, POSSIBLY, PAID SERVICE. SUCH A VERSION IS POSSIBLE.

((UGLANOV)) AND, FINALLY, RUMORS, PERHAPS, HAVE REACHED ME TO THE EFFECT THAT UP UNTIL RECENTLY A NUMBER OF TOP SCIENTISTS HAD BEEN ENLISTED IN A STUDY OF UFO'S AT THE DISPOSAL OF THE MINISTRY OF DEFENSE. I AM TALKING ABOUT SOME FRAGMENTS OR WHOLE "ARTICLES," WHICH ARE BEING KEPT SECRET IN HANGARS AND UNDERGOING TESTS.

((KOTENKOV)) WORK IS BEING PERFORMED ON THE GLOBAL MONITORING OF SPACE AND THE AIR SPACE ON THE TERRITORY OF THE FORMER SOVIET UNION. ALL UNIDENTIFIED FLYING OBJECTS ARE RECORDED. ANY OBJECT ENTERING THE FIELD OF VISION OF OUR EARLY WARNING STATIONS IS, OF COURSE, MONITORED. BUT I HAVE NOT HEARD OF ANY OBJECTS BEING STORED AND STUDIED ANYWHERE.

ADMIN
(ENDALL) ███████ 21/1720Z MAY
BT

#7466

NNNN

Figure 62•C00179737 Top Russian scientists enlisted in study of UFOs

C00015255

A

July 1965

PUNITIVE STEPS AGAINST EXTREMIST ASKED

Havana PRENSA LATINA in Spanish 2140 GMT 5 July 1965--E (FOR OFFICIAL USE ONLY)

(Excerpt) Buenos Aires, 5 July--While rumors of a coup are increasing, the government, in a McCarthy-like campaign launched in recent weeks by several reactionary groups, has just taken the first official step toward repressing the leftwing forces by means of a resolution which lumps together the fascist organizations and the Communist Youth Federation. Interior Minister Juan Palmero issued a resolution which demands that the Justice Ministry take punitive measures against the members and supporters of the Tacuara nationalist movement, its Nationalist Students Union branch, the Nationalist Restoration Guard, and the Communist Youth Federation (PJC).

According to a special report issued by the federal police upon the request of the ministry, the resolution is based upon the charge that the organizations are acting as "veritable lawless associations" in "carrying out violent actions." It is also planned to instruct the federal police to give the utmost cooperation to the Justice Department in order to investigate the facts and, when necessary, to arrest the perpetrators.

- 0 -

MEATPACKING INDUSTRY CRISIS--Buenos Aires--Two of the largest meatpacking plants in Argentina, Swift and Armour, have reduced their operations, claiming that the Argentine currency they receive for their imports does not cover the costs of meat prices in the domestic market. Numerous workers have been fired as a result of this reduction in operations, thus compounding the meatpacking crisis. (Buenos Aires ANSA Spanish 0126 GMT 6 July 1965--P) Sources close to the Argentine economic cabinet have reported that the restriction of meat consumption will ban the handling and consumption of meat on Fridays and Saturdays instead of Mondays and Tuesdays. Exports will be stepped up, exempting steers for export from taxes. The economy minister has received a committee representing the 8,000 meatpacking industry workers affected by a decline in operations, assuring them that the government is considering measures to reactivate those industries to promote exports of meat and its byproducts. (Buenos Aires ANSA Spanish 1556 GMT 6 July 1965--P) (FOR OFFICIAL USE ONLY)

RELATIONS WITH ALGERIA--The Argentine Foreign Ministry has unofficially announced that Argentina will continue to have diplomatic relations with Algeria. An official communique will be issued in a few days. (Buenos Aires ANSA Spanish 1556 GMT 6 July 1965--P) (FOR OFFICIAL USE ONLY)

ANTARCTIC FLYING SAUCERS--A group of red, green, and yellow flying saucers has been seen flying over Deception Island for two hours by Argentine, Chilean, and British bases in Antarctica. The flying saucers were also seen flying in formation over the South Orkney islands in quick circles. (Buenos Aires ANSA Spanish 1556 GMT 6 July 1965--P) (FOR OFFICIAL USE ONLY)

CRUDE OIL IMPORTS- It has been learned that the Argentine YPF has announced that bids from private companies will be received for the purchase of 1 million cubic meters of crude oil. (Lima AFP Spanish 1835 GMT 6 July 1965--P) (FOR OFFICIAL USE ONLY)

T-4

Approved for Release
2/2010

Figure 63•C00015255 Antarctic UFO sighting

C00386418

```
*** Document 199 of 54 ..: FBIS ***
DOCN 000103001
CLAS UNCLAS 3A/PMU
SERI SERIAL:    AU3003152893
PASS PASS:      ATTN BBC SD
COUN COUNTRY:   RUSSIA INTERNATIONAL
SUBJ*SUBJ:      PAPER REPORTS ALLEGED EVIDENCE ON MISHAP INVOLVING UFO
SOUR SOURCE:    KIEV HOLOS UKRAYINY IN UKRAINIAN 27 MAR 93 P 5
TEXT TEXT:
        //((REPRINT FROM THE NEWSPAPER TERNOPIL VECHIRNIY:  "COSMIC
     REVENGE" -- FIRST PARAGRAPH PUBLISHED IN BOLDFACE))
        ((TEXT))  AFTER MIKHAIL GORBACHEV DISSOLVED, IN 1991, THE KGB TOP
     SECRET INTELLIGENCE ADMINISTRATION, A LOT OF MATERIAL FROM THAT
     DEPARTMENT FOUND THEIR WAY ABROAD, IN PARTICULAR TO THE CIA.  AS
     REPORTED BY THE AUTHORITATIVE MAGAZINE CANADIAN WEEKLY WORLD NEWS,
    *U.S. INTELLIGENCE OBTAINED A 250-PAGE FILE ON THE ATTACK BY A UFO ON
     A MILITARY UNIT IN SIBERIA.
        THE FILE CONTAINS NOT ONLY MANY DOCUMENTARY PHOTOGRAPHS AND
     DRAWINGS, BUT ALSO TESTIMONIES BY ACTUAL PARTICIPANTS IN THE EVENTS.
     ONE OF THE CIA REPRESENTATIVES REFERRED TO THIS CASE AS "A HORRIFIC
    *PICTURE OF REVENGE ON THE PART OF EXTRATERRESTRIAL CREATURES, A
     PICTURE THAT MAKES ONE'S BLOOD FREEZE."
        ACCORDING TO THE KGB MATERIALS, A QUITE LOW-FLYING SPACESHIP IN
     THE SHAPE OF A SAUCER APPEARED ABOVE A MILITARY UNIT THAT WAS
     CONDUCTING ROUTINE TRAINING MANEUVERS.  FOR UNKNOWN REASONS,
     SOMEBODY UNEXPECTEDLY LAUNCHED A SURFACE-TO-AIR MISSILE AND HIT THE
    *UFO.  IT FELL TO EARTH NOT FAR AWAY, AND FIVE SHORT HUMANOIDS WITH
     "LARGE HEADS AND LARGE BLACK EYES" EMERGED FROM IT.
        IT IS STATED IN THE TESTIMONIES BY THE TWO SOLDIERS WHO REMAINED
     ALIVE THAT, AFTER FREEING THEMSELVES FROM THE DEBRIS, THE ALIENS
     CAME CLOSE TOGETHER AND THEN "MERGED INTO A SINGLE OBJECT THAT
     ACQUIRED A SPHERICAL SHAPE."  THAT OBJECT BEGAN TO BUZZ AND HISS
     SHARPLY, AND THEN BECAME BRILLIANT WHITE.  IN A FEW SECONDS, THE
     SPHERES GREW MUCH BIGGER AND EXPLODED BY FLARING UP WITH AN
     EXTREMELY BRIGHT LIGHT.  AT THAT VERY INSTANT, 23 SOLDIERS WHO HAD
     WATCHED THE PHENOMENON TURNED INTO... STONE POLES.  ONLY TWO
     SOLDIERS WHO STOOD IN THE SHADE AND WERE LESS EXPOSED TO THE
     LUMINOUS EXPLOSION SURVIVED.
    *   THE KGB REPORT GOES ON TO SAY THAT THE REMAINS OF THE UFO AND THE
     "PETRIFIED SOLDIERS" WERE TRANSFERRED TO A SECRET SCIENTIFIC
     RESEARCH INSTITUTION NEAR MOSCOW.  SPECIALISTS ASSUME THAT A SOURCE
     OF ENERGY THAT IS STILL UNKNOWN TO EARTHLINGS INSTANTLY CHANGED THE
     STRUCTURE OF THE SOLDIERS' LIVING ORGANISMS, HAVING TRANSFORMED IT
     INTO A SUBSTANCE WHOSE MOLECULAR COMPOSITION IS NO DIFFERENT FROM
     THAT OF LIMESTONE.
        A CIA REPRESENTATIVE STATED:  "IF THE KGB FILE CORRESPONDS TO
     REALITY, THIS IS AN EXTREMELY MENACING CASE.  THE ALIENS POSSESS
     SUCH WEAPONS AND TECHNOLOGY THAT GO BEYOND ALL OUR ASSUMPTIONS.
     THEY CAN STAND UP FOR THEMSELVES IF ATTACKED."
     (ENDALL)     23003.03 27 MAR                     30/1529Z MAR
     BT
     #0317
NNNN NNNN
---EOD---
```

Approved for Release
Date

MAY 2000

(6)(3)

39

Figure 64 •C00386418 KGB in possession of remains of UFO

214

C00015467

CENTRAL INTELLIGENCE AGENCY

INFORMATION FROM
FOREIGN DOCUMENTS OR RADIO BROADCASTS

REPORT NO. OO-W-44108

CD NO. --

COUNTRY	Algeria; Spain; French Equatorial Africa	DATE OF INFORMATION	1952
SUBJECT	Military - Unconventional aircraft		
HOW PUBLISHED	Daily and thrice-weekly newspapers	DATE DIST. 27 Sep 1952	
WHERE PUBLISHED	Tangier; Conakry; Oran; Casablanca	NO. OF PAGES 4	
DATE PUBLISHED	22 Apr - 1 Aug 1952		
LANGUAGE	Spanish; French	SUPPLEMENT TO REPORT NO.	

THIS IS UNEVALUATED INFORMATION

SOURCE   Newspapers as indicated.

"SAUCERS" SIGHTED OVER SPAIN AND FRENCH AFRICA

GLOWING SPHERES SEEN OVER ALMANSA, SPAIN -- Tangier, Espana, 22 Apr 52

Almansa, 21 April -- At 0800 hours today, many persons saw a series of four glowing spheres crossing the sky along the Murcia-Valencia trajectory at high speed and at a great height. The spheres looked like locks of wool of a vivid reddish color which changed to an intense yellow as they flew farther away. They gradually disappeared in the distance. No unusual sound was heard. The spectacle lasted hardly a minute.

ANDUJAR RESIDENTS REPORT SAUCER -- Tangier, Espana, 1 Aug 52

The Cifra news agency reports that on the night of 30 July 1952, many residents of Andujar, Spain, saw what was presumed to be a so-called flying saucer. The object was red, round, and approximately the size of a desert dish. It flew noiselessly, at great speed, and left behind a long trail of very bright greenish light.

LUMINOUS OBJECT SEEN OVER PORT GENTIL -- Conakry, La Guinee Francaise, 26 Jun 52

The master of a cargo ship anchored in front of the wharf of Port Gentil [in Gabon, French Equatorial Africa] reported that at 0240 hours on 1 June 1952 a mysterious object came up from the area behind Port Gentil, made a double loop, passed over the roadstead, and then dived toward the sea at great speed. He submitted the report to the local authorities and to the administrative offices of his company. The following is a summary of his story.

On 1 June 1952 at 0240 hours, the ship was riding at anchor in the roadstead of Port Gentil, heading 150 degrees. The northern sky was clear and starry; the southern, slightly cloudy. Visibility was excellent, a slight southwest breeze prevailed, and the sea was calm. There was a quarter moon.

ARCHIVAL RECORD
PLEASE RETURN TO
AGENCY ARCHIVES,

- 1 -

6 nov. 78

Figure 65 •C00015467 UFO sighting from cargo ship page 1 of 2

C00015467

The first mate was at his forward post ready to weigh anchor while the master was on the bridge with the officer on duty. With the exception of the mooring lights, the ship was in complete darkness, thus permitting excellent night visibility.

At 0240, the first mate telephoned the master informing him that he had just sighted an unknown luminous object in the sky which came from Port Gentil and passed directly over the ship. Training his binoculars (Zeiss 7 x 50, for night vision) skyward, the master was able to see quite clearly, on the port quarter, a very bright and phosphorescent orange light, circular in shape and moving at a great speed in a seemingly straight-line course. Standing on the wing of the bridge, and with the aid of the gyrocompass repeater, the master estimated its average direction to be about 10 degrees.

He followed the light quite easily in his binoculars for about 3 minutes and lost sight of it when it moved at great speed over the Prince buoy, about 7 miles from the ship. The master was unaware of any accompanying sound and admits that it was difficult to estimate the altitude of the object, yet he judged this to be 3,000-4,000 meters. Its diameter was that of a planet.

The first mate stated that before he telephoned the master, he saw that object come from the direction of Port Gentil, stop, make a right turn, and resume its initial course. As it passed directly over the ship, it repeated the same sort of gyration.

The master stated that his 20 years of sea duty enabled him to affirm that what he saw was neither a known celestial phenomenon, such as a falling star or meteor, nor a current type of aircraft.

Furthermore, it was confirmed that there were no planes in the air that night over Port Gentil.

STRANGE OBJECTS SEEN IN SKY OVER ALGERIA -- Oran L'Echo d'Oran, 17 Jul 52

At 2300 hours on 15 July 1952, in the town of Boukanefis, two bakers clearly saw a plate-shaped flying object in the sky. It moved with unusual agility, giving off a greenish smoke and lighting up the sky. It seemingly did not alter its course as it increased its speed and disappeared toward the south.

Similarly, in Lamoriciere on 11 July 1952, one Thomas Martinez saw a sudden illumination above and at first took it to be a falling star. Actually, it seemed more like a meteor followed by two other bodies, all trailing a yellow cloud of smoke. Then, these disappeared and out of nowhere appeared an oval-shaped, longish ball of fire. Flying at a low altitude and clearly visible, it rapidly followed a rectilinear course.

STRANGE PHENOMENON WITNESSED FROM TWO POINTS IN ORAN -- Oran, L'Echo d'Oran, 26 Jul 52

Recently /presumably 25 July 1952/, several inhabitants of Oran reported seeing a flying saucer at approximately the same time.

At 1535 hours, Raoul Le Henaff, foreman of a local company, saw an incandescent white mass in the sky above Oran. It flew southwestward, traveling at an altitude of about 2,000 meters. There was no luminous exhaust and no smoke. After 30 seconds, the phenomenon grew hazy and disappeared. The personnel of the company office and Palacio, an employee in the Algeronaphte /factory?/ also saw the "saucer."

- 2 -

Figure 66•C00015467 UFO sighting from cargo ship page 2 of 2

216

C00112351   UNCLASSIFIED

UNCLASSIFIED

                                                              PAGE:0020

COPY TO ELAAD, ESG, MOD (2)

BODY
SUBJ

BARBADIANS REPORT SEVERAL UFO SIGHTINGS

FL021725

BRIDGETOWN CANA IN ENGLISH 1427 GMT 2 SEP 87

      ((TEXT)) BRIDGETOWN, BARBADOS, SEPT 2, CANA -- SCORES OF
BARBADIANS LAST NIGHT REPORTED SPOTTING SEVERAL UNIDENTIFIED FLYING
OBJECTS IN THE MOONLIT SKY AND PRESS REPORTS TODAY SAID THE GLOWING
BALLS WERE ALSO SEEN ELSEWHERE IN THE EASTERN CARIBBEAN.
      EYE WITNESSES ON THE GROUND HERE SPOKE OF SEEING BETWEEN FOUR AND
18 SLOW-MOVING BALLS OF LIGHT WITH LONG ILLUMINATED TAILS, GLIDING
HORIZONTALLY THROUGH THE SKY IN A NORTH-SOUTH DIRECTION OVER THIS
166-SQUARE-MILE ISLAND. THE SIGHTINGS LASTED ABOUT 10 MINUTES, SOME
SAY.
      METEOROLOGICAL OFFICIALS COULD NOT EXPLAIN THE OBJECTS, BUT THE
PRESS QUOTED A PILOT OF THE REGIONAL AIRLINE LIAT AS SAYING THAT
WHILE FLYING AT ABOUT 8000 FEET, A GROUP OF OBJECTS HAD PASSED HIS
AIRCRAFT "AT A TERRIFIC SPEED" -- A LARGE ONE OUT FRONT, FOLLOWED BY
ABOUT FIVE OR SIX OTHERS.
      THE MET OFFICE SAID SIMILAR SIGHTINGS HAD ALSO BEEN REPORTED IN
GRENADA, ST LUCIA, AND MARTINIQUE.
ADMIN
(ENDALL) 021427███████████████02/2008Z SEP
BT
#0356

          NNNN

                         UNCLASSIFIED

Figure 67 •C00112351 1987 Barbados UFO sighting

C00386416

```
*** Document 305 of 54 ..: FBIS ***
DOCN 000429524
CLAS UNCLAS 6A
SERI SERIAL:   FL2312181593
PASS ATTN:     ATTN BBC SD
               COPY TO ████
COUN COUNTRY:  CUBA
SUBJ*SUBJ:     UNIDENTIFIED FLYING OBJECTS SIGHTED IN MATANZAS SKIES
SOUR SOURCE:   HAVANA RADIO PROGRESO NETWORK IN SPANISH 1200 GMT 23 DEC 93
TEXT TEXT:
        //((REPORT BY CANDIDO DOMINGUEZ FROM MATANZAS))
        ((TEXT)) THE DAY BEFORE YESTERDAY, A GROUP OF MATANZAS RESIDENTS,
   *YOURS TRULY INCLUDED, HAD THE OPPORTUNITY TO OBSERVE UNIDENTIFIED
   *FLYING OBJECTS, UFO'S, OVER MATANZAS CITY. SHORTLY BEFORE 1900, I
   HEARD MY EIGHT-YEAR-OLD DAUGHTER WHO WAS ON THE PORCH SHOUT: DADDY,
   LOOK, FIREWORKS. I WALKED OUT ONTO THE PORCH AND OBSERVED THE
   STRANGE PHENOMENON. THE FIRST THING I SAW WAS A VERY BRIGHT WHITE-
   BALL FOLLOWED BY LIGHTED TAIL, SOMETHING LIKE A COMET. LATER THE
   LIGHTS MULTIPLIED AND I SAW SIX OR SEVEN OF THEM LINED UP, EACH WITH
   A TAIL THAT FOLLOWED A NORTH-TO-SOUTH TRAJECTORY. THESE LIGHTS
   REMAINED VISIBLE FOR APPROXIMATELY 30 SECONDS.
        YESTERDAY I LEARNED THAT A GROUP OF SPECIALISTS FROM THE
   TERRITORIAL METEOROLOGY SECTION OF THE CUBAN ACADEMY OF SCIENCES
   ALSO VIEWED THE STRANGE SIGHT FROM VARIOUS AREAS OF THE CITY. ONE OF
   THEM, JESUS GOMEZ, SAW IT IN JOVELLANOS, 42 KM SOUTHEAST OF
   MATANZAS.
        (ORESTES GRIBAO), WHO IS A METEOROLOGIST AND FOR YEARS HAS BEEN
   *STUDYING UFO'S, WATCHED THE PHENOMENON IN THE VERSAILLES
   NEIGHBORHOOD. (GRIBAO) SHOWED SEVERAL PEOPLE A BOOK THAT HAS A
   PHOTOGRAPH TAKEN IN 1913 FROM THE TORONTO OBSERVATORY, CANADA. WE
   ALL AGREED THAT WHAT WE SAW ON 21 DECEMBER IN MATANZAS IS ALMOST
   IDENTICAL TO WHAT THE 60-YEAR-OLD PICTURE SHOWS. UNDOUBTEDLY, THIS
   WILL BE A SUBJECT OF MUCH INTEREST TO YTHOSE WHO STUDY THE
   *EXTRATERRESTRIAL UNIVERSE.
        (ENDALL) 231200 ████████    23/1815Z DEC  ██222312.013
   BT
   #0906
NNNN NNNN
---EOD---
```

Approved for Release
Date

MAY 2008

Figure 68•C00386416 1993 CUBA UFO sighting

C00779653

*by UFO second time in one year

Text of report by Radio Russia on 24 August

*   Unidentified flying objects have again been spotted in the sky
above Tyumen Region's Nizhniy Tavda District. For two nights in a
row they could be seen above the District's capital and the nearby
lakes and rivers.
    The RIA news agency reports that this is this year's second
*appearance of UFOs above Nizhniy Tavda. Last time an unidentified
*flying object was witnessed by local residents on 1 January. Then a
huge sphere was seen hovering above the settlement [of Nizhniy
Tavda] and sending eight rays of light towards the ground.
*   This time, the UFO presented itself in two forms: that of a
sphere and a saucer. In addition, local residents are claiming that
they saw an almost 100-m-long aircraft with two evenly glowing
sidelights. The flying object was moving very slowly and without
producing any noise whatsoever.
    The residents of Nizhniy Tavda are now trying to fathom what is
*it that drives UFOs to their small, quiet and not in any way
distinguished settlement buried in the Siberian sticks.

    (endall)
    BT
    #3174

NNNN NNNN
PPER (407) SERIAL NOT FOUND OR INCORRECTLY FORMATTED.
ELST   407 - NONCRITICAL (PPER.L):
            SERIAL NOT FOUND OR INCORRECTLY FORMATTED.
EFLG NONCRITICALERROR
===EOD===

Figure 69 • C00779653 Russian radio report of UFOs over area lakes and rivers

C00779655

"research bases" in Caspian Sea

    Text of unattributed report by Azerbaijani newspaper 525 qazet
*on 25 August entitled "A flying saucer appears over the Caspian
again"

*   Unidentified flying objects have become more active over the
Caspian Sea. Last week, residents of some villages near Baku
sighted alien objects again. Some villages even had blackouts when
*UFOs flew over the Caspian Sea.
    According to information given to Olaylar news agency by Fuad
Qasimov, head of the seismology department of the Azerbaijani
*National Aerospace Agency, UFOs have research bases in the Caspian.
It is possible to observe them mainly over the villages of Pirsaqi,
*Nardaran and Suvalan. Apart from the Caspian, UFOs have bases in
the Black Sea, the Mediterranean, Iran, Africa and in several
American states. Qasimov said that the latest event was filmed on
amateur video.

    THIS REPORT MAY CONTAIN COPYRIGHTED MATERIAL.   COPYING AND
DISSEMINATION IS PROHIBITED WITHOUT PERMISSION OF THE COPYRIGHT
OWNERS.
(endall)
BT
#5276

NNNN NNNN
PPER (407) SERIAL NOT FOUND OR INCORRECTLY FORMATTED.
ELST   407 - NONCRITICAL (PPER.L):
            SERIAL NOT FOUND OR INCORRECTLY FORMATTED.
EFLG NONCRITICALERROR
===EOD===

Figure 70 •C00779655 UFOs over Caspian Sea

Project Sun Streak Remote Viewing Project

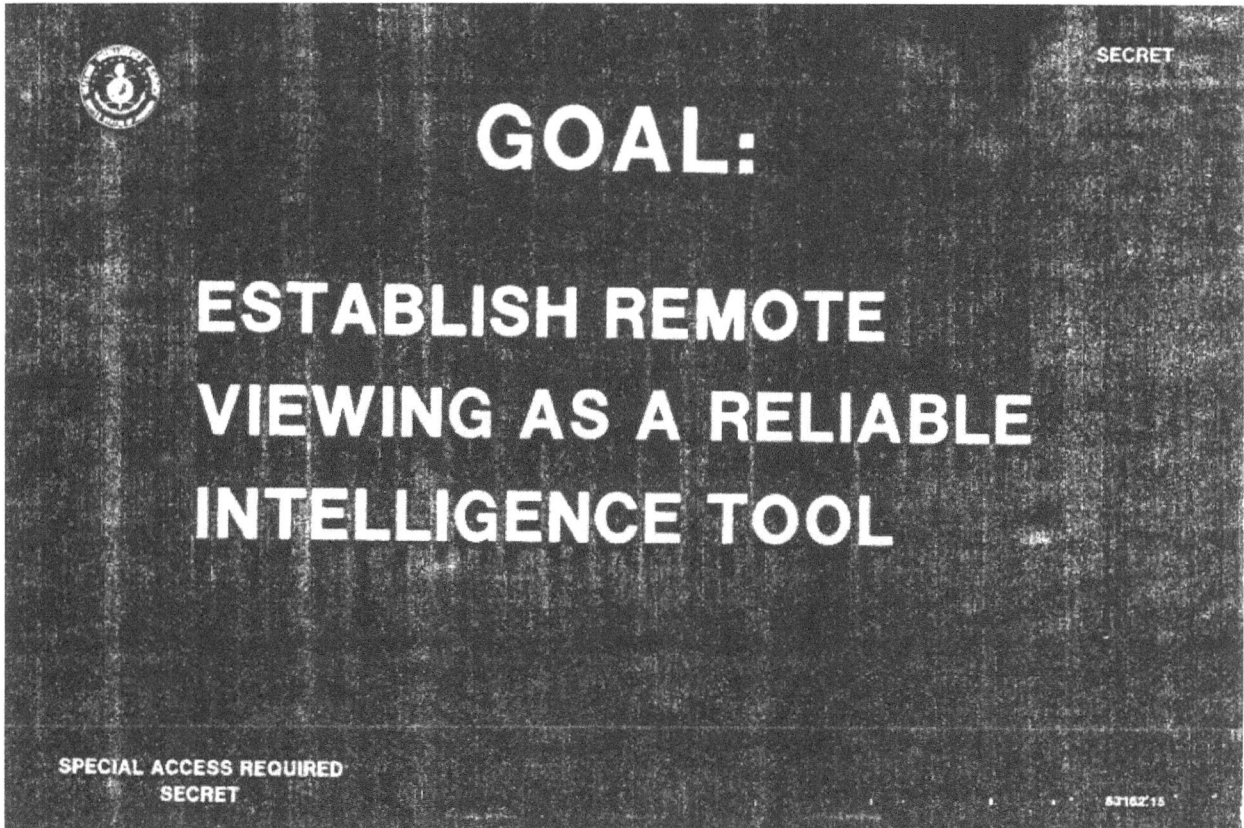

Debbie Ziegelmeyer

DEFENSE INTELLIGENCE AGENCY

WASHINGTON, D.C. 20340-6150

S-20004/DT-S   1 1 AUG 1987

MEMORANDUM FOR THE DEPUTY SECRETARY OF DEFENSE

SUBJECT: Request for renewal of human use approval - ACTION MEMORANDUM (U)

(S/NF/SS-2) This request is in response to your desire for annual review of the DIA psychoenergetics activity conducted under SUN STREAK, a DoD-approved special access program.

(S/NF/SS-2) During FY 87, the 11 people assigned to Project SUN STREAK pursued the practical applications of applied psychoenergetics against selected operational tasks and improved skills through proven training techniques.

(S/NF/SS-2) During FY 88, proven training techniques will be continued and a wider range of operational tasks will be pursued for cueing Intelligence Community sources in the collection of foreign intelligence.

(S/NF/SS-2) The SUN STREAK Project Tasking Group comprised of a panel of senior executives from DIA, CIA, AIA, AF/IN, DNI and C^3I oversees and advises in matters relating to operational concerns. Project activity is monitored and reviewed by an executive-level panel which includes $ASDC^3I$ and DR/DIA.

(S/NF/SS-2) Request you reapprove use of DoD personnel for psychoenergetics activity in the SUN STREAK program conducted with no changes in program procedures and in strict compliance with the provisions of Procedure 13, DoD directive 5240.1-R.

LEONARD H. PERROOTS
Lieutenant General, USAF
Director

Coordination:

OSD (GC) by H. L. Garrett, III

APPROVED

OTHER

SG1J

Prepared by

CLASSIFIED BY: DIA/DT
DECLASSIFY ON: OADR

WARNING NOTICE: This material is restricted to those with verified access to SUN STREAK LEVEL 2 (SS-2).

DEFENSE INTELLIGENCE AGENCY
WASHINGTON, D.C. 20340-0001

19 January 1990

U-022/DT

SG1J

TO: ▮▮▮▮▮▮▮▮
DR-XO

SUBJECT: Sun Streak

Frank:

Enclosed is the memo we discussed on Sun Streak   SG1I
to implement the MIB decision. ▮▮▮▮▮▮▮agrees
with it, except he wants to have OER/appraisal
signature authority. Per our personnel folks, we
can't do that since they're DIA people still at
Ft Meade. This lets him review the OERs and
appraisals though, so it's workable.

Regards,

SG1J

Enclosure a/s

Vice Assistant Dep Dir
for Scientific and
Technical Intelligence

Debbie Ziegelmeyer

UNITED STATES GOVERNMENT

memorandum

DATE:

REPLY TO
ATTN OF: DR

SUBJECT: Sun Streak Program

THRU: VP
TO: DT

SG1I

1.   In accordance with a recent MIB decision, operational control of the subject program has been transferred to the Army Staff. ████████████████████ DAMI-PO, is the focal point on the program. All inquiries on the program should be referred to ████████████

SG1I

SG1J

2.   Intermittent DT support for the program, in the person of ████████ ██████ may be required for the next few months. Please establish the necessary procedures to have ████████████ available for periodic consultations as necessary, consistent with his normal workload.

SG1J

SG1J

SG1I

SG1I

3.   The program staff will remain assigned to DIA for the duration of the test program this fiscal year. The manpower authorizations are and will remain in DIA. While operational control of the staff will be exercised by ████████████ administrative control and support will remain with DT, including payroll accounting, and appraisals and OER closeout. ████████████ is authorized to monitor appraisal and OER preparation and will coordinate their closeout through DIA/DT; no reporting official changes will be made at this time as a result of this decision. Any requirements for TDY by the program staff will be funded by the Army project during this period. Within DIA, DT will be the focal point for program accesses, and will coordinate accesses with ████████████

SG1I

cc:
DAMI
DAMI-PO
RHR
OC

OPTIONAL FORM NO. 10
(REV. 1-80)
GSA FPMR (41 CFR) 101-11.6

☆ U.S. GOVERNMENT PRINTING OFFICE : 1982 O - 361-526 (8017)

C00212889

DOCN 000573950
FBIS FBIS UNNUMBERED   FOR OFFICIAL USE ONLY
CLAS UNCLAS

SUBJ ████████████████████████████

///have psychotic breaks."

UK: U.S. Use of 'Psychic Spies' Reported (Take 4 of 5)

SERI LD2708141095

TEXT   [FBIS Transcribed Text] Lt P- recovered, and remains on active
duty but Stubblebine retired from the Army in 1984 to become an
executive at BDM Corporation a Washington-area defence and
intelligence contractor. He left BDM a few years ago, and now lives
in New York, where he is married to Rima Laibow, a controversial
psychiatrist who has claimed that she is a UFO abduction victim. But
the damage had been done. "Bert gave remote-viewing a bad name,
because of all the other stuff he was involved in," says a former
senior Pentagon official who knew him. And although the unit never
left its offices at Fort Meade, by 1986 it had been expelled from
the Army. It still had its supporters, notably Jack Vorona, chief
of the DIA's science and technology directorate, who had since 1978
been the overall head of the remote-viewing programme. The DIA took
the Fort Meade unit under its wing, the project was renamed Center
Lane, and later, Sun Streak, and Vorona now exerted more direct
control of the Fort Meade unit. For the remote-viewers, this was a
fortunate development. Vorona was a man who was widely respected
throughout the intelligence community, and with him watching over
it, the unit seemed safe from outside threats.
        But what of inside threats? Although Stubblebine was gone, his
spirit lingered, and in the mid and late 1980s, the unit seemed to
take on a garish tinge. In its first few years under DIA management
the unit included the "witches,", two women called Angela Dellafiora
and Robin Dahlgren. Dellafiora eschewed remote-viewing and instead
"channelled" her psychic data through a group of entities with names
like "Maurice" and "George". Dahlgren practiced tarot-card reading.
        In the eyes of Ed Dames and Mel Riley, Angela achieved an undue
influence on the unit when she began to give personal channelling
sessions, featuring advice on the most intimate matters of their
lives, to Jack Vorona and other officials. "Jack Vorona would sit at
one end of the table, and Angela at the other," recalls Dames. "She
would say, 'Good morning, Dr Vorona. Maurice says hello!'"
        "Their eyes would be shining when they came out of those
sessions," recalls Riley. "They were told all the nice things they
wanted to hear, which reinforced Angela's position within the unit."
        "Psychic blowjobs," says Ed Dames, referring to the activities of
Angela and Robin. To witness them, he told me, and the other antics
of "the witches", was "too much to bear for professional military
officers". But Dames as much as anyone was caught up in the
transformational dynamic of remote-viewing.
        A linguist - his forte was Chinese - and former INSCOM
intelligence officer, Ed Dames was one of the group that had been
trained in the early Eighties by Ingo Swann at SRI. With his blond
hair, California accent, and preternaturally boyish face, he looked
more like a teenage surfer than a soldier. Although widely
considered intelligent and creative, he also seemed, like

Approved for Release
2/2010

Stubblebine, to have an impulsive streak. "Everybody sort of looked
at Ed as a loose cannon," says Mel Riley. "I was in trouble all the
time, anywhere I went," agrees Dames. "I was always pushing the
envelope."
        Certainly, despite his professed distaste for the New Ageishness
of Vorona and the "Witches", Dames was frustrated by the increasing

C00212889

scarcity of operational taskings. In his ample spare time at the
unit, he began to use remote-viewing techniques to exercise his own
*spiritual and extraterrestrial interests. "Under the guise of
'advanced training,'" he says, "I began to see what [remote-viewing]
could do. You know what I mean?" Dames's advanced training
"targets" included apparitions of the Virgin Mary, the demise of
Atlantis ("it's at the bottom of Lake Titicaca," says Dames), the
*Loch Ness monster ("a dinosaur's ghost"), and a great many flying
*saucers. "He would tell me a lot of things about Martians,"
remembers Dames's now estranged wife Christine. "I didn't want to
hear about it."

While Dames was at the Fort Meade unit, stories began to
circulate about certain "unusual experiences" during remote-viewing
sessions, particularly those engaged on "advanced training" targets.
"I think he had some kind of experiences, some kind of disturbances
from unknown spirits," remembers Christine Dames. "But he didn't
care -- he welcomed the challenge."

"We thrived on adventure," Dames remembers proudly. "You get men
of action -- we're not satisfied with sitting around and twiddling
our thumbs year after year," says Dames. "Unless something happens,
you're going to lose our interest. But there was enough happening
in there to hold our interest."

Dames left the unit in 1989, and formed a company, Psi Tech, to
make commercial use of his remote-viewing skills. But his clients
were few and far between. He separated from his wife and moved to
Albuquerque, New Mexico, believing that the nearby deserts harboured
a hidden Martian civilisation. A wilderness prophet for our time,
he predicted to the local media that in August 1992, the aliens
would arise from their desert dwellings, shocking the world. When I
saw him in 1994, Ed Dames was almost out of money.

MOST OF the remote-viewers I've talked to are willing to admit,
when pressed, that their craft does have its psychiatric hazards. As
with any prolonged and forced alteration of consciousness, it
promotes altered states and a general mental instability, and thus
can be dangerous for those who are inherently unstable. They also
point out that in the absence of regular independent verification,
remote-viewing can quickly become a generator of idiosyncratic
fantasy. As Mel Riley says, "Without feedback, your remote-viewing
turns to shit."

And without proper oversight, it seems, the remote-viewing
programme turned foul, too, slowly strangled by its own isolation.

Following the Irangate scandal of 1987, Defense Secretary Frank
Carlucci had instituted a wide-ranging review of potentially
embarrassing Pentagon programs, and in 1988, a Defense Department
Inspector General's (IG) team descended on the remote-viewing unit's
offices, demanding to see the files.

(more)

27 AUG 1610z

NNNN NNNN
===EOD===

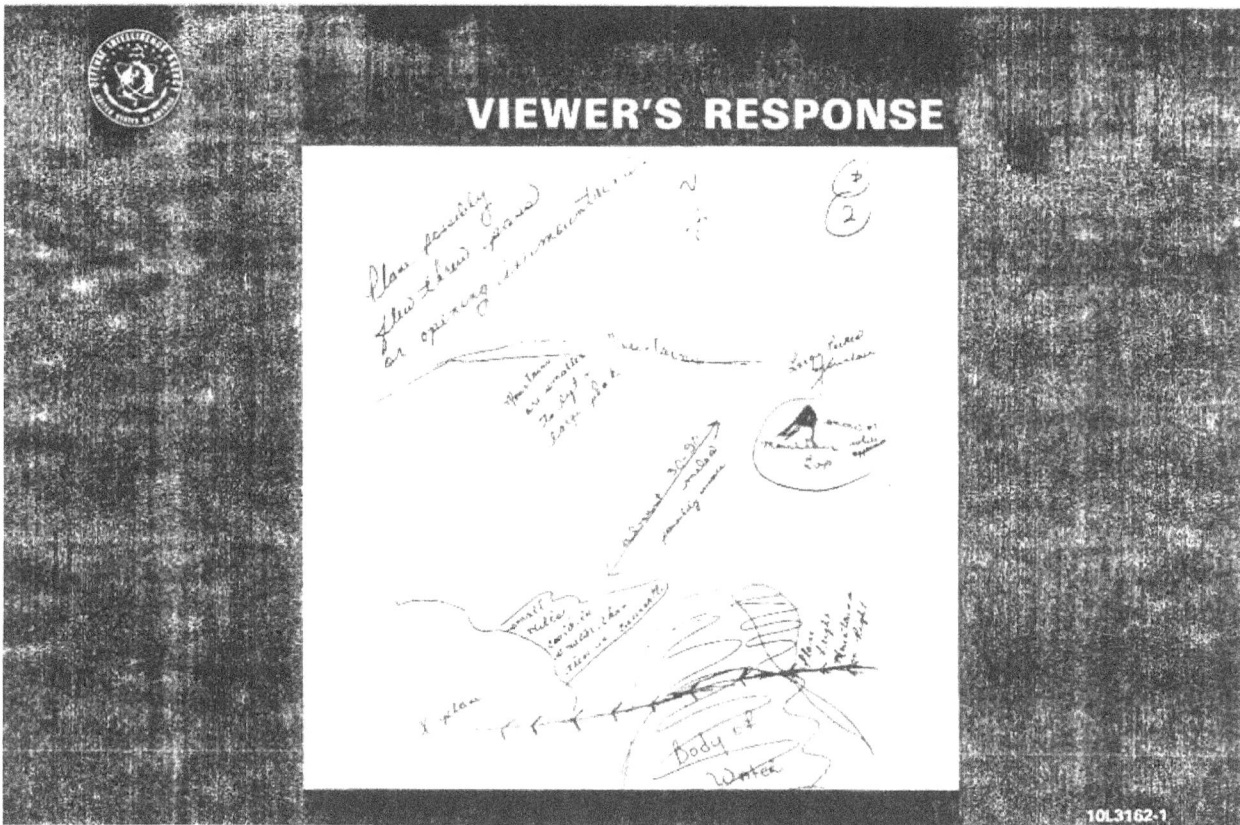

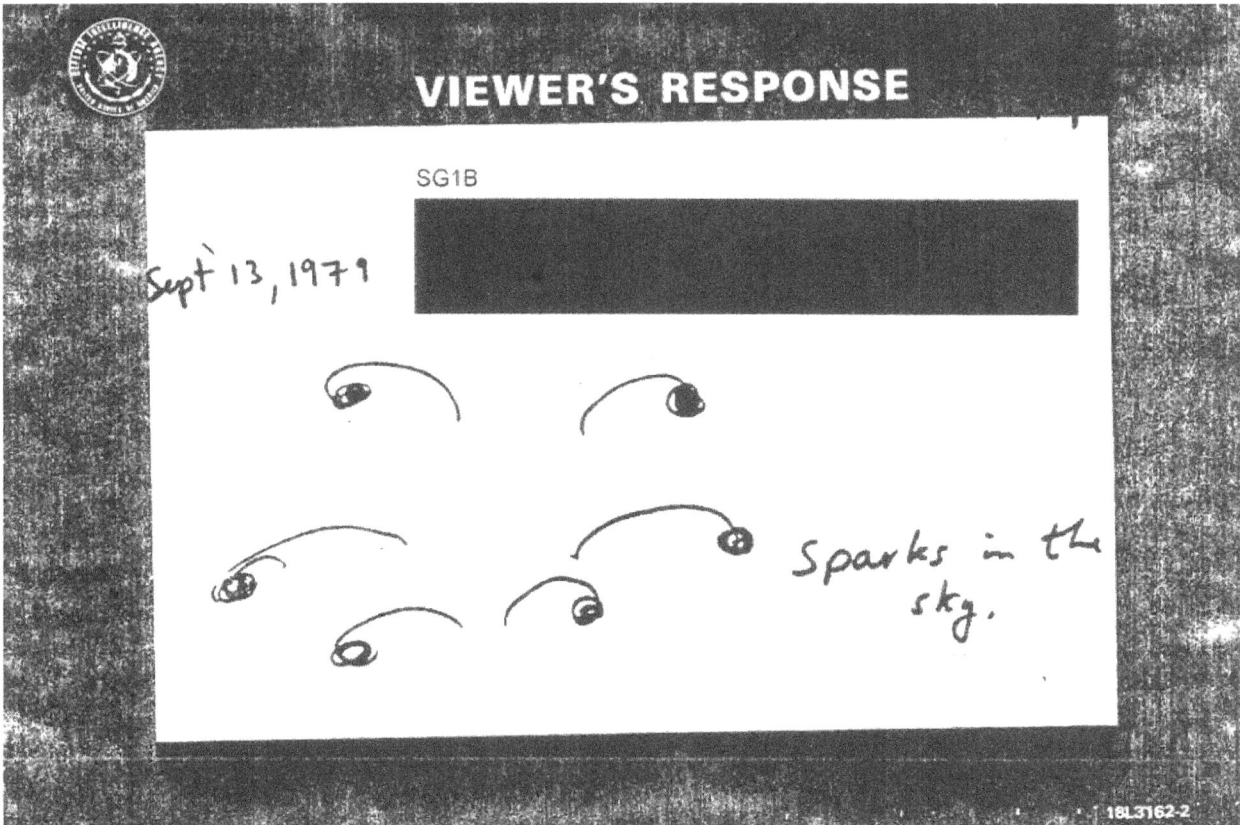

Debbie Ziegelmeyer

~~SECRET/NOFORN SKEET CHANNELS ONLY~~

PROJECT <u>SUN STREAK</u> (U)

<u>WARNING NOTICE: INTELLIGENCE SOURCES AND METHODS INVOLVED</u>

CRV Session Procedures Report (S/NF/SK)

```
┌─────────────────────────────────┐
│Control Number  : 8704           │
│                                 │
│Date of Session : 08Apr87        │
│                                 │
│Date of Report  : 08Apr87        │
│                                 │
│Source Identifier: 021           │
└─────────────────────────────────┘
```

   1. (S/NF/SK) Tasking: See original tasking data sheet, attached.
No other instructions received.

   2. (S/NF/SK) Session: Began with a review of source's last session
(1 April 87) and a discussion of possible avenues of approach to the as
yet unanswered question of what the "large structure" is all about.
Monitor suggested that source consider earlier-perceived, target-
associated concepts of "opaque" and "self-contained" and to allow some
freedom of movement in time. Source decided to begin in S4 with the
self-cue of "structure". Monitor has a high level of confidence with
regard to this session. [Note: Interestingly, at one point in the
session both source and monitor had the same thought regarding this
site, to wit: "What ever happened to the simple scientific-technical
facility " (Source) and "These kinds of things are starting to become
out of our ball park" (Monitor)].

   3. (S/NF/SK) Summary: A solid, oddly-shaped geometric (see source
sketch) object is located inside the structure. There is a pulsing,
greenish light/energy associated with it. A high-frequency hum is
present nearby. The object is somehow involved with the idea (process)
of attracting something like "particles" (AOL-"fairy dust") from outside
the large structure, which seems to cause the associated pulsing and
warm, iridescent green light/energy. These "particles" seem to
originate from space.

   4. (S/NF/SK) Feedback: None provided source.

SG1J

CPT, USA

~~SECRET/NOFORN SKEET CHANNELS ONLY~~

CLASSIFIED BY: DIA-DT
DECLASSIFY   : OADR

light — there is warmth to it.

Some of the particles is
outside structure —
there's like a "fan"
beam of these coming
from above. It is as
though the structure
has been placed
to collect this.

Structure contains a solid geometric object

which collects/attracts particle - like "dust" from a
very distant dark source away from the
earth and converts it into a pulsing
irridescent green warm energy. There is
a low-level frequency/pitched humming
associated with the object. No sense of people
present.

Conclusion

After researching the existence of underwater "Alien" bases, I found numerous locations of "best possibilities" for both underwater bases and another means such as vortex or wormhole entry. In my decades of research on the subject of the alien colonization of our planet, evidence leads to a strong possibility that some bases are human/alien occupied under government treaties. I also found through research of witness reports and descriptions of the sightings, that not all alien life-forms visiting, or colonizing our planet make their way here in the same fashion. We learned from the document in Chapter 11: "Blue Planet Project", that there are a supposed 160 different alien life forms who have visited the Earth over several centuries. Even if the number is inaccurate, we know from countless witness reports and descriptions, that numerous different alien life-forms exist and have been or are currently on our planet. One cannot assume that these file-forms all act and react in a similar manor, nor can we assume they travel to and from their place of origin in the exact same manor.

In Chapter 4: Mysterious Triangles, research pinpoints areas where EMF (electromagnetic fields) exist to the extent that the areas effect the weather, ocean wave patterns, vessels which dare to transverse these areas and private, commercial, and military aircraft which fly above. Documents such as "The Aeronautical Information Manual (AIM)" issued for the area of Lake Michigan, and Bermuda Complex; a warning posted for the Bermuda Triangle area on the GlobalSecurity.org website, reflect the dangers of crossing these mysterious areas. There are numerous anomalous zones on our planet that seem to be consistent locations for missing persons and mysterious disappearances of vessels and aircraft. Some of those who transverse these areas simply vanish, never to be found. Could there be a possibility that these mysterious "High-EMF" zones are locations where "alien life-forms" travel to and from our planet by means of vortexes, wormholes, time travel, or dimensional travel? Are missing people subjects of research and study by "alien life-forms" or victims of being "in the wrong place at the wrong time"? This is a mystery still to be solved as well as a "high interest" area for UFO researchers to follow.

Years of research, witness reports, and released government documents lead UFO researchers to the conclusion that "alien life-forms" reside either permanently or temporarily in facilities located in mountains, deep underground, and in bases located deep underwater. My research took me across the world to reports of multiple sightings of "unidentifiable craft" possibly "alien", hovering over, going into, coming out of, and moving at high rates of speed beneath the Earth's oceans, seas, lakes, and rivers. Although recent video has been released by the U.S. government of some of these incidents such as the "Nimitz video", unidentified vessels and aircraft have been documented or entered into sailor's logbooks for centuries. Governments worldwide have been denying the existence of "other-worldly life" since the 1940's. The only difference today by releasing documents and captured video is a so called "playing it down" strategy. The governments of the world especially in the U.S., are telling us what many UFO investigators have known for over 75 years, that "alien life forms" are traveling to Earth from other planets or moons, are here on this planet, and have been for centuries. As one family member of a high-ranking U.S. politician recently stated when he was asked directly "How are you guys getting away with this? Like, aren't you concerned?" The gentleman looked up, laughed a little bit, and replied, "Plausible deniability". Although the lies being told were of a different subject matter the premise is the same.

There is sufficient evidence to make a legitimate case of the alien colonization of our planet Earth. A strong possibility of "alien life-form" presence in these waterways exists.

Pacific Ocean
Alaska
Catalina-Channel Islands
Hawaiian Archipelago
Long Beach, California
Moonless Mountains
Monterey Bay, California
Puget Sound Region Pacific NW
San Diego and Channel Islands
Sea of Japan

Atlantic Ocean
Andros Island, Bahamas (AUTEC)
East Coast of North America
Guantanamo Bay
Kings Bay, Georgia
Mayport, Florida Naval Station
Norfolk, Virginia
Puerto Rico

Other Areas of Interest
Antarctica
The Arctic
Baltic Sea
Great Lakes United States
Great Britain
Hainan Island, China
Kola Peninsula, Russia
Lake Baikal, Southern Siberia
Lake Titicaca, Peru-Bolivia
Mediterranean Sea

North Island, New Zealand
North Sea Coast
Northwest Coast of Scotland
Puffin Island, Wales
Red Sea
Solomon Islands

There are multiple documents available discussing underwater human occupied bases and facilities being planned and built as early as the 1960s. Several researchers-investigators such as Vladimir Azhazha, Dr. Virgilio Sanchez-Ocejo, Lt. Col. Wendelle C. Stevens, Preston Dennett, Carl W. Feindt, Maximillien de Lafayette, Jorge Martin, Ivan Sanderson, Richard Sauder, Phillip Mantle, Paul Stone, and many others including myself, have accumulated thousands of hours of research, documents, witness interviews, police and military reports, photos and videos documenting the existence of not only underwater military bases but the possibility of "alien being bases". Volunteer organizations such as MUFON, NUFORC, CUFOS, "The Black Vault" and many others past and present, have also dedicated themselves to uncovering the truth behind the UFO phenomenon.

The UFO phenomenon does not limit itself to "strange unidentified craft" or "mysterious unidentified lights" but involves humans on a personal level. Several witness reports are of abductees being taken underwater to dry base-like facilities. Many abductees tell of personal violations through physical examinations or worse. Others, such as in "Triangle" areas, simply vanish, never to be seen again.

Any subject matter involving the field of UFO investigation is based on evidence available and speculation to fill in the gaps. My documentation has covered a history of human underwater research, witness experiences, and military involvement past and present. It also includes water related UFO sighting reports some dating back hundreds of years, long before humans were safely capable of navigating beneath our oceans, seas, and waterways. The probability of underwater "alien life" bases or habitats is a very strong possibility based on this research and that of many others in the field of ufology.

The following questions need to be considered:

- Are alien races not only visiting but colonizing our planet and have they been doing so for centuries?

- Is there enough evidence available to prove the existence of underwater alien bases or alien vortex-wormhole hole doorways?

- Are alien beings harvesting Hydrogen from the Earth's water sources, and why?

- Have humans across the world, been abducted for centuries by beings from other worlds?

- Are world leaders aware of the alien presence?

- Are these world leaders working scientifically with some of these extraterrestrial races?

From recent events across the world, citizens have come to realize that they are and have been lied to for decades by the government officials they elected, appointed, and recruited. There is one truth for you the public and another for the privileged few who hold the world's deep dark secrets. All we can do is forge ahead in an effort to uncover as much of the truth as possible, relying on those in government who believe the public has a right to know its fate whatever it may be, and are willing to leak out bits and pieces of the truth.

For every question asked there are seemingly twelve answers and for each of those twelve answers given there are an additional twelve questions to be asked.......

References

"10 Alleged Underwater Alien UFO Bases" Marcus Lowth June 13, 2016

"Amazing Soviet-Era UFO Sightings In & Over Water Bodies"

AmericanCivilWarStory.com

ancient-origins.net

Bible: St. James Version

"Blue Planet Project: Alien Techical Research 25"

Catalina Islander July 8, 1947

chazzsongsufos.

Cryptopia

"Did the US Navy Build Super Secret Undersea Lairs for its Nuclear Submarines?"

Steve Weintz/TweetShareShare/The National Interest August 6, 2016

Dailymail.co.uk

express.co.uk/news/science

EMN earthmysterynews.com

haarp.gi.alaska.edu

Filer's Files

Fox News

gendisasters.com

Google Earth

"Hidden in Plain Sight" Richard Sauder, PH.D.

 https://www.shipwreckmuseum.com/

"Is there a massive underwater UFO base in Guantanamo Bay?" Stacy Liberatore for Dailymail.com

"Inside UFOs" Preston Dennett

islapedia.com

"Invisible Residents" Ivan T. Sanderson

 "James Cameron Now at Ocean's Deepest Point" Ker Than, National Geographic News

Jean-Michel Cousteau's Ocean Futures Society

jewishpress.com

Jorge Martin

Jules Undersea Lodge Greatdreams.com

Kevin Loria Florida International University/Aquarius

Linda Porter

Los Angeles Times

messagetoeagle.com

military.wikia.org

MUFON archives

mufon.com

National Geographic

NASA

NASA Earth Data

NASA/Kim Shiflett

natgeotv.com

Navy.com

neveralone51.blogspot.com

Debbie Ziegelmeyer

NOAA
Northern Divers USA
NW Rocketeer 02-16-73
Nwo Report1
PADI
Paul Stonehill Rense.com 10-23-00
Phillip Mantel
Pinterest.com
Publicado por LUJÁN ARCHIVOS OVNI
Politico
Popular Mechanics
Popular Science
"Russia's USO Secrets" Paul Stine & Philip Mantle
Russian Ministry of Defense / Barents Observer
Roane State Community College
SCU Scientific Coalition for Ufology
Scubadiverlife.com
Seaward Marine
"Shining the Light" Light Technology Research
shipwreckmuseum.com
Soviet Merchant Ship UFO Sighting notes: Robert Morningstar/testimony: Alexander G. Globa
The Anomalist "UFOs in Soviet Waters" Paul Stonehill
The Black Vault
The Debrief
The Russian Ufology Research Center
"The Underwater cases: Russian Underwater Encounters"
"UFO Abduction From Undersea" Dr. Virgilio Sanchez-Ocejo/Lt. Col. Wendelle C. Stevens (Ret.)
"UFO Case Files of Russia" Paul Stonehill and Phillip Mantle
"UFO Commentary" Bill Wickersham, ED.D.
"UFO-USO and Extraerrestrials of the Sea" Maximillien de Lafayette
"UFOs and Water" Carl W. Feindt
"UFO Undersea" Dr. Virgilio Sanchez-Ocejo/Lt. Col. Wendelle C. Stevens (Ret.)
"Undersea Base-AUTEC) " greatdreams.com
.ufoinsight.com
"Underwater Anomalies: In the Andes of Peru" creatorsdream
10 Alleged Underwater Alien UFO Bases Marcus Lowth June 13, 2016
"Underwater and Underground Bases" Richard Sauder, PH.D.
"Underwater Bases and Alien Civilization" Altantis Rising: Library
"Under-Ocean Military Bases: There Is More Than Just One Area 51" Arjun Walia January 28, 2016
 Underwater Construction Corporation
"Unlimited Impossibilites: Intelligence Support to the Deepwater Horizon Response"
nsf.gov
Capt. Erich M. Telfer, USCG
Vladimir Azhazha
"Vieques Caribbean UFO Cover-up of the Third Kind" Jorge Martin
Washington Post
Wikipedia
Wikileaks
worldatlas.com
yesterday.uktv.co.uk

New York Times
Zagadki Sfinksa magazine (issue #3, 1992) odessa Translation by Paul Stonehill

INDEX

About the Author

Debbie Ziegelmeyer of Imperial, Missouri, was born in Grand Rapids, Michigan, attended school in Safford, Arizona and later attended high school first in Southern California, then Arnold, Missouri where she graduated from Fox Senior High School.

Debbie joined MUFON in 2000 and became a Field Investigator shortly after. She is currently a MUFON Star Team Investigator, the State Director of Missouri, a member of the MUFON Business Board of Directors, The Functional Director of Underwater Research and Recovery, a Benefactor, and Inner Circle member, and a MUFON Archivist. She is also a member of Ted Phillip's S.I.U./T.P.R.C. Team, and in 2004, organized the MUFON Dive Team along with Tom Ferrario current T.P.R.C. Team leader. Debbie has over 1200 completed UFO investigations, with

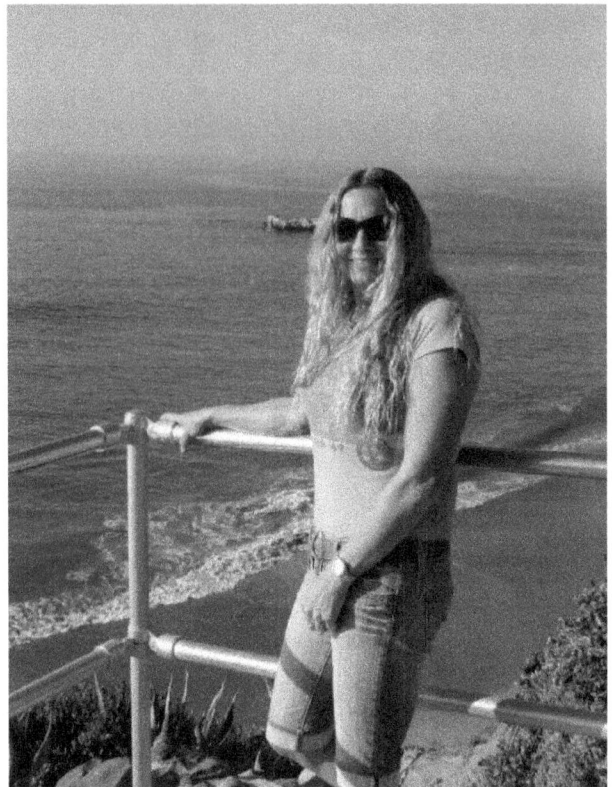

Figure 71 Debbie on the California Coastline

over 860 of them being MUFON sighting reports. She also teaches a MUFON Field Investigator Dive Team course which instructs investigators in both in water and on shore responsibilities and protocols.

Debbie has been a Roswell 1947 crash investigator for over 25 years, has been an annual Roswell, NM July Festival speaker, and has made several speaking appearances at the "UFO Watch Tower" in Hooper, Colorado. She was a volunteer archeologist on two NBC projects, 2002 "Roswell Crash Starling New Evidence", which published her personal experiences in the book "Roswell Dig Diaries", and the 2006 UFO Hunters episode "Roswell Revisited". She appeared on an episode of "Close Encounters" as an expert on the "1970s UFO Flap in Piedmont, Missouri," and was an expert consultant on a San Luis Valley episode of "Mysteries in National Parks" on the Travel Channel. Debbie was a 2017 Devils Tower "UFO Rendezvous" speaker, a 2018 "Alien Cosmic Expo" speaker in Toronto, Canada, a 2013 and 2017 MUFON Symposium speaker, a 2019 MUFON Cruise speaker, an Arkansas Paranormal Expo speaker in 2018, 2019 and 2021 in Little Rock Arkansas, and a 2021 OZ Conference speaker. She has also made several personal speaking appearances in Missouri has been a guest SKYPE and ZOOM speaker and been a guest on several radio shows and podcasts.

Most recently, Debbie appeared on two episodes of "Alien Highway" as a consultant, two episodes of "Alaska Triangle", and the 2021 Discovery+ series "Roswell the Final Verdict". Debbie is also featured in the Ben Mezrich book "The 37th Parallel" which is a chronological account of her brother Chuck

Zukowski's UFO and cattle mutilation investigations. In 2017 she was awarded both the "Roswell Incident Excellence Award" by the Roswell Daily Record, and MUFON "Field Investigator of the Year".

Debbie's full-time occupation is small family business owner-manager of Largo Properties, LLC. and Custom RV & Boat Storage. Debbie became a PADI Scuba Dive Instructor in 2007 and holds teaching credentials in several dive instructor specialties including divers with special needs. She currently serves on the Board of Directors of "World of Water Foundation," a 501c3 dedicated to giving disabled veterans and children with special needs, the chance to experience scuba diving and snorkeling.

Publications by Un-X Media

Rules for Goddesses by Margie Kay 1999

Haunted Independence Missouri by Margie Kay 2013 & 2016

Gateway to the Dead: A Ghost Hunter's Field Guide by Margie Kay 2016

Family Secrets by Jean Walker 2017

The Kansas City UFO Flaps by Margie Kay 2017

A Sonoma County Phenomenon: Evidence for an Interdimensional Gateway by Margie Kay 2019

The Fast Movers: Evidence for High-Speed UFOs/UAPs

by Margie Kay, Bill Spicer, and Larry Tyree 2020

Journey to Spirit by Devin Listrom 2020

Winged Aliens by Margie Kay 2021

The Alien Colonization of Earth's Waterways by Debbie Ziegelmeyer 2021

The Master Dowsers Chart Book by Margie Kay 2022

Missouri UFO Hot Spot by Missouri MUFON 2022

THOR by Margie Kay 2022

All books available at www.amazon.com and www.barnesandnoble.com

UnX News Magazine

Available in print or PDF at magcloud.com

UnX News Blog

unxnews.blospot.com

UnX News Radio and Podcast

www.unxmedia.com

Un-X Media is currently taking book submissions. We publish non-fiction books about unexplained phenomena. Please check the website for writer guidelines.

Contact:

editor@unxmedia.com

816-833-1602

www.unxmedia.com

UNXMEDIA

PUBLISHING

www.ingramcontent.com/pod-product-compliance
Lightning Source LLC
Chambersburg PA
CBHW080328270326
41927CB00014B/3133